Smart Innovation, Systems and Technologies

Volume 261

The Smart Innovation, Systems and Technologies book series encompasses the topics of knowledge, intelligence, innovation and sustainability. The aim of the series is to make available a platform for the publication of books on all aspects of single and multi-disciplinary research on these themes in order to make the latest results available in a readily-accessible form. Volumes on interdisciplinary research combining two or more of these areas is particularly sought.

The series covers systems and paradigms that employ knowledge and intelligence in a broad sense. Its scope is systems having embedded knowledge and intelligence, which may be applied to the solution of world problems in industry, the environment and the community. It also focusses on the knowledge-transfer methodologies and innovation strategies employed to make this happen effectively. The combination of intelligent systems tools and a broad range of applications introduces a need for a synergy of disciplines from science, technology, business and the humanities. The series will include conference proceedings, edited collections, monographs, handbooks, reference books, and other relevant types of book in areas of science and technology where smart systems and technologies can offer innovative solutions.

High quality content is an essential feature for all book proposals accepted for the series. It is expected that editors of all accepted volumes will ensure that contributions are subjected to an appropriate level of reviewing process and adhere to KES quality principles.

Indexed by SCOPUS, EI Compendex, INSPEC, WTI Frankfurt eG, zbMATH, Japanese Science and Technology Agency (JST), SCImago, DBLP.

All books published in the series are submitted for consideration in Web of Science.

More information about this series at https://link.springer.com/bookseries/8767

Elspeth McKay
Editor

Manage Your Own Learning Analytics

Implement a Rasch Modelling Approach

Springer

Editor
Elspeth McKay
Cogniware.com.au
Melbourne, VIC, Australia

ISSN 2190-3018 ISSN 2190-3026 (electronic)
Smart Innovation, Systems and Technologies
ISBN 978-3-030-86315-9 ISBN 978-3-030-86316-6 (eBook)
https://doi.org/10.1007/978-3-030-86316-6

This Springer imprint is published by the registered company Springer Nature Switzerland AG
The registered company address is: Gewerbestrasse 11, 6330 Cham, Switzerland

Preface

Overview

There is a broad awareness of data analytics across educational institutions for various stakeholders. The way learning analytics is defined involves: assessment reports for individual learners to know how they rate compared with other learners; to highlight students who may need extra support; to assist teachers planning their supporting interventions for individuals and groups of learners; for professional development teams when considering new courseware design and development; and for institutional/corporate marketing and recruitment management strategies. However, the practice of managing learning analytics for some people may seem a daunting task. This preface uses a six-step prescriptive learning analytics planning model to guide the novice instructional system designer to show why this perception is wrong. This model will equip people with the information they need to carry out learning analytics, including: careful preparation of their testing instruments; developing an understanding of the importance of validating the measurement tools; recognising a poorly designed instrument; and interpreting the output files.

Objectives of the Book

Manage Your Own Learning Analytics aims to contribute to the Springer book series *Smart Innovation, Systems and Technologies* (series editors: R. J. Howlett and L. C. Jain). The book's main objective is to tease out the range of learning analytics techniques used by the global research community and enable interested researchers to publish their learning analytics methodologies, no matter their methodological design or processes used to develop the validity and reliability of instruments to measure (knowledge) constructs used as variables in their research. For example, Rasch modelling and the item response theory (IRT) embraced by researchers worldwide are especially popular in the behavioural sciences. Nevertheless, it appears that the business sector is yet to discover this powerful means of drilling into data sets to initiate fine-grained analyses. The following chapters represent a range of approaches dealing with learning analytics. The chapters were all peer-reviewed, offering authors' suggestions for improving their proposals; chapters were rejected with a consensus of review opinion and when the topics did not align with the book's intent. Thus, this book is an opportunity to showcase learning analytics, regardless of the methodology perspective.

Target Audience

Industry training developers, corporate trainers, courseware designers, government sector specialists, educational technology practitioners and researchers, lecturers/teachers (higher education, primary and secondary) and postgraduate students are the target audience.

Learning Analytics in Six Easy Steps

Implementing learning analytics takes various forms according to need, such as: compiling assessment reports for individual learners to know how they compare with other learners; highlight students who may need extra support; assist teachers in planning interventions for individuals and groups of learners; support professional development teams when considering new courseware design and development; and support institutional/corporate marketing and recruitment management strategies (McKay, 2019). However, for some people, assessing how much people learn or understanding the instructional programme's effectiveness may seem beyond their scope of experience. Instead, by following a prescriptive learning analytics planning model, in which time and energy spent on matching task objectives to required knowledge levels and careful preparation of their assessment instruments, people should be able to carry out their own learning analytics (McKay, 2019).

Step-1: Prepare test instrument Think about the overall learning objectives, and choose a gradual skill-building or task progression matrix. For example: start with the easy concepts or declarative knowledge *(knowing that);* through mid-range tasks *(intellectual skills)*; to initiating procedural or cognitive strategies *(knowing the how)* Gagné matrix (McKay, 2000 p. 175).

Design a test specification (skill-building) instrument specification matrix that depicts two separate pedagogical functions to determine skill/knowledge achievement levels. First, conduct a thorough task analysis. List the steps (starting with the easiest ending with the most difficult) (Theng, 2012) needed to achieve the learning objectives for each task (vertical axis, Fig. 1), and determine the types

Instructional Objectives : Making a Pizza

		Declarative		Procedural			
		Band-A	Band-B	Band-C	Band-D	Band-E	
		Verbal information skill	Intellectual skill	Intellectual skill	Cognitive Strategy	Cognitive Strategy	
		Concrete concept	Basic rule	Higher-order-rule	Identify sub-tasks	Knowing the 'how'	
		Knows basic terms	Discriminates	Problem solving	Recognizes unstated assumptions	Recall simple prerequisite rules & concepts	
		Knows 'that'	Understands concepts & principles	Applies concepts & principles to new situations		Integrates learning fromdifferent areas into a plan for solving a problem	
Task No:	Learning Domain:						Totals:
6	Make sauce					2 questions	2
5	Make dough				3questions		3
4	Use oven			2 questions			2
3	Measure ingredients			2 questions			2
2	Read recipe	1 question	2 questions				3
1	Decode abbreviations	2 questions					2
	Totals:	3 questions	2 questions	4 questions	3 questions	2 questions	14

Fig. 1 Test instrument specification matrix (adapted from Mat-Jizat, 2012: Mager, 1988)

of declarative and procedural knowledge development expected for each task (horizontal axis, Fig. 1). Second, write out test-items according to where they plot on the matrix. A well-designed skill-level test will show test-items as a gradual skill-building progression (Theng, 2012).

Step-2: Set scoring regime Dichotomous (yes/no answers), multiple-choice and polytomous (stepped scoring—Likert-type items). First, write out acceptable answers in preparation for marking scheme development; then, allocate scoring for each test-item.

Step-3: Validate testing instrument Assessment and measurements concerning intended trait/knowledge construct are super important factors to consider (Andrich & Marais, 2019); arrange to have your test-items checked by experts.

Use an appropriate software application to check that your test-items are a fit for the Rasch model (Bond & Fox, 2015), for example, the Windows-based Rasch Unidimensional Measurement Model (RUMM2030) www.rummlab.com.au; Conquest 4 (Windows) www.acer.edu.au/conquest; Winsteps and Facets (Windows) www.winsteps.com/facets.htm; and WINMIRA (Windows) https://winmira-2001.software.informer.com/1.0.

Enter the scored test outcomes into the Rasch measurement application (usually submitting a test score input file, often as a spreadsheet or text file). Then, run the application, and examine the result. In some Rasch measurement software applications, you may need to remove test-items considered bad questions from the input file; rerun your item analysis until all test-items are a Rasch model fit (Fig. 2). However,

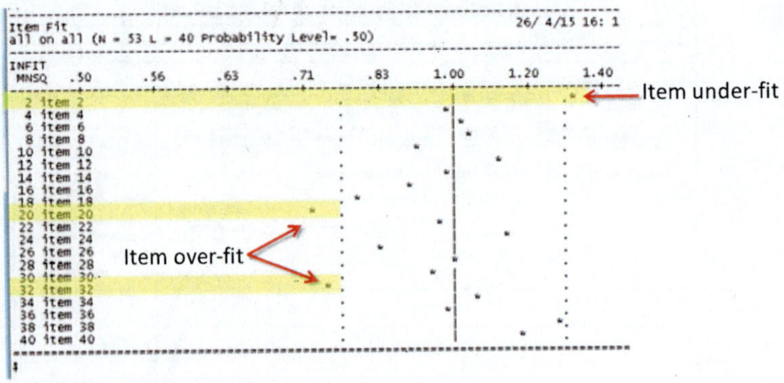

Fig. 2 Rasch model—item fit (adapted from Bakkar, 2016)

it is unnecessary when using the RUMM2030 application; instead, such data can be removed or readjusted to fit the model's requirements (Robinson, Johnson, Walton, & MacDermid, 2019).

Figure 2 represents Rasch estimate data shown here as a data map from an earlier Rasch measurement tool, QUEST Interactive Test Analysis System (Adams & Khoo, 1996). The vertical dotted lines represent the fit thresholds; items to the right of the upper threshold (1.25) underfit the Rasch model and are considered bad items that must be removed from the test score input file, while test-items to the left of the lower threshold (.74) overfit the Rasch model, so are redundant items that can also be removed from the input file.

These Rasch measurement applications provide a unidimensional scale, with equal intervals (Fig. 3), that measures people's performance (each X on the left-hand side represents one participant) and test-items together (numbered on the right-hand side).

Often referred to as a Wright Map or person-item map, Fig. 3 highlights the very complex nature concerning rating-scale data and test data. It shows the persons' ability and is relative to all other persons, relative to test-items behaviour and relative to behaviour of all other items, on a common unidimensional/logit scale. Three people are shown here as an X on the left side of this map, with a mean ability estimate of 1.3, while the test-items are captured on the right side (laid out according to the 5-Likert scale intervals—1. never been used; 2. not effective; 3. effective to some extent; 4. effective; and 5. very effective). Although these maps are automated by the Rash model software, constructing them by hand provides a deeper understanding (Boone, Staver & Yale, 2014).

Therefore, this person/item variable map conveniently illustrates the different Likert-scaled agreement levels, for each question—located on the unidimensional scale. It should be noted to look into this map more closely to see that often the third level of the question has a higher logit value than the level-4 of that same question, which may be an indication of participants selecting the mid-point of the Likert scale, if they were not sure of their response level.

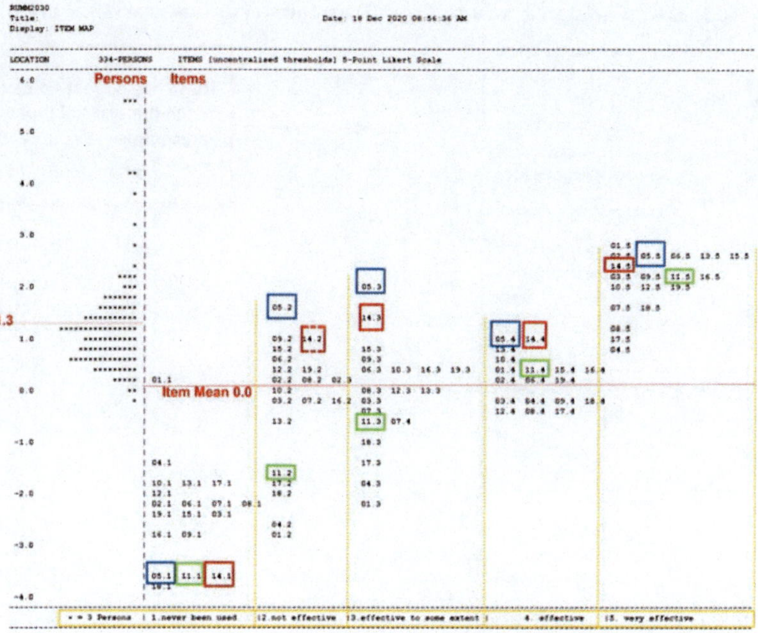

Fig. 3 Rasch model variable map sometimes called a Wright Map (Boone, Staver, & Yale, 2014)

Step-4: Modify test-items Remove test-items considered poorly written questions (only when your Rasch measurement tool requires this. See earlier mention of RUMM2030 capabilities (Robinson, Johnson, Walton, & MacDermid, 2019), in Step-3.

Check the compatibility of the model and the data through the item fit statistics in the Rasch measurement application printouts, deleting or modifying non-fitting test-items as they shift along the scale throughout this process. This iterative process has very powerful benefits, such as revealing what can happen without careful attention to non-fitting test-items. For example, Fig. 4 depicts a poorly designed instrument that was too easy for the students/trainees.

Step-5: Implement test Ensure all questions are answered to avoid zero scores.

Properly validated test-items can be given to participants (students/trainees) in a pre- and post-test assessment instrument. For instance, when investigating the effectiveness of an instructional strategy/learning programme, a pretest will determine the level of skills/knowledge before people undertake the programme, while the post-test will measure any change/knowledge acquisition after the instructional intervention.

Step-6: Analyse results
A key focus of an educational researcher is to design reliable test instruments for measuring performance outcomes in particular settings. Adopting the Rasch

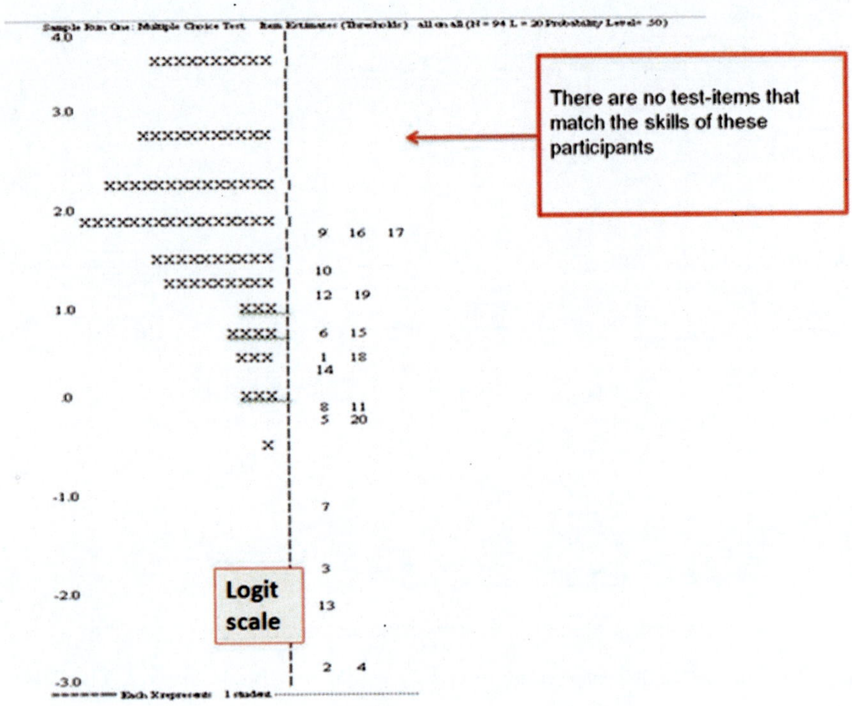

Fig. 4 Rasch model variable map showing a poorly designed instrument

measurement approach assures control of the analysis and interpretation of data. In this sense, the knowledge constructs' test score reliability property (or behaviour) reflects how consistent or error-free the measurements were (Frisbie, 1988). Such test results can show whether the scores achieved represent whether learning has occurred and whether the instructional strategies were effective for that particular student cohort. In addition, it is helpful to understand student reactions to the test-items. For instance: consistently low scoring across the cohort may indicate the need for remedial instructional strategies (for particular students), or where test-items were too easy for others, requiring additional problem-solving activities/test-items to stretch those students cognitively.

Expressing the magnitude of change in a student/trainee's proficiency following an instructional programme, as the magnitude or size of effect, as defined by Cohen's statistical power analysis (Cohen, 1977), was popular with researchers using the QUEST Interactive Test Analysis System (Bakkar, 2016). In addition, some Rasch model applications provide a QUEST item analysis output table (Fig. 5). This table offers the best of both worlds because it gives both measurement practices of classical test theory (CTT) and item response theory (IRT) in establishing the Rasch model's discrimination value.

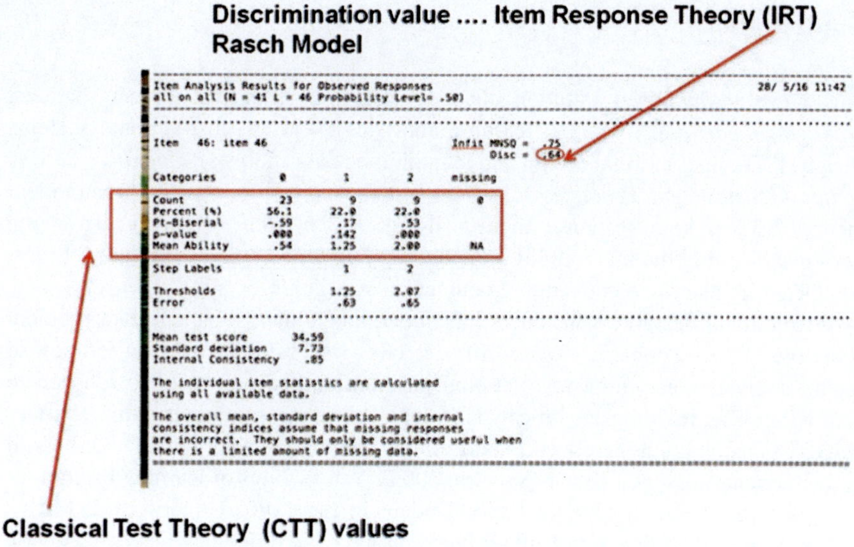

Fig. 5 Test-item analysis table

Following these simple six steps will allow novice educators/trainers to develop their own reliable and validated learning analytics, thereby understanding if their learning materials achieve the expected result from the learning/training cohort, if not, where improvements should be made.

Scholarly Value and Contribution

The chapters in this book directly compare approaches to learning analytics across many philosophical fields and vary according to personal researcher perspectives and organisational visionary and cultural positions. The global nature of the chapter authorship represents research from six countries: Australia; Malaysia; The Netherlands; Poland; Saudi Arabia; and South Africa), revealing the impact of increased awareness for promoting valid and reliable learning analytics in diverse professional practices, including: information systems; social sciences, media, film and education; access health and community; creative industry management and performing arts; higher educational education; online learning; and machine learning.

Contributions

This book is organised into ten chapters and falls into four topical sections: *Managing Learning Analytics* (learning analytics overview, instructional systems design (ISD), instructional design and planning the data analysis); *Cognitive Performance Measurement Practices* (classical test theory (CTT), Rasch measurement theory (RMT), item response theory (IRT)); *Rasch Modelling Tools* (research design and setting the methodology); and *Case Studies* (corporate training settings, healthcare industry and educational courseware design).

Instructional design practitioners have been aware of assessment measurement since the 1970s (Popham, 1978; Griffin & Nix, 1991) due to the proliferation of digital analysis concerning instructional programme evaluations (also referred to in this book as learning analytics). However, when massive open online courses (MOOCs) came upon the educational/training scene, they were touted as killers of higher education as we knew it (Spector, 2005). Yet, feelings of learning limitations for online academic superstores were evident in the 1990s (Meerilees & Miller, 1997). Therefore amidst differing opinions on what learning analytics is, this book is concerned with translating how to understand the personal learning needs and performance outcomes (Greller & Drachsler, 2012) from a range of perspectives.

Privacy-Driven Learning Analytics—This opening chapter provides an excellent overview of the book. The authors explain that data privacy is a central theme in the global dialogue around the application of data science in education. However, despite the growing need, research organisations and private companies working on education and learning analytics solutions still rely on ad hoc, red tape heavy and inconsistent approaches to privacy protection. This chapter outlines the substantial and growing body of work on data privacy risk measurement and reduction, which can help address this problem and better use online learning data with improved privacy risk management.

Using Log Stream Data and Item Response Theory to Understand Proficiency in Collaboration—This chapter provides an exemplar approach to convey the nature of skills such as collaboration is complex. There are internal processes at play, and inferences need to be made to interpret explicit behaviours observed from intentionally designed assessment tasks. The theoretical concept of the paths presented in this chapter is to a priori define the processes being sought in the process stream data and interpret their substantive meaning concerning the skill measured. This chapter centres on an analytical learning approach to reduce complex skills into more straightforward but observable components that can be represented as aspects of the target construct and scaled to understand the growth of such skills. The method outlined in this chapter presents an approach that transforms process stream data of collaboration into formative assessment data that can be analysed using item response theory. The application of this model allows the interpretation of the data as levels of proficiency that we can use to map or monitor progress in collaborative skills.

Introductory Analysis of the Rasch Model—This chapter is written primarily for novice learning analytics researchers wishing to understand the Rasch measurement approach. The author's encounter with the Rasch model came after a long experience teaching traditional statistical methods, then being called in to advise a Ph.D. student on test-items used for developing tests: the Rasch measurement theory (RMT). This technique develops efficient psychometric tests that produce interval measures. This approach is an astonishing feat, and it took the author some time to recognise that this was based solidly on probability theory, particularly regression. This experience, plus becoming involved in the Ph.D. supervision, led to an investigation of the ideas behind the RMT, its 'philosophy' and the process of calibration, particularly the use of fitness measures. This kind of introduction to Rasch is certainly not rare. Furthermore, statistics is traditionally a hurdle for many students, so perhaps the explanations given here benefit from the teachers' perspective, even if potentially suffering because of it. The text, therefore, relates elements from probability theory. It also explains the fundamental aspects of psychometric testing, which is rarely covered in basic statistical texts. This chapter also provides a critique of significance and sample size related to RMT and, to some degree, to regression techniques in general. This subtle matter is perhaps appreciated by many, but it is briefly analysed here.

Data Evaluation of Happiness Scale Online Study: A Rasch Measurement Analysis—This chapter represents a collaborative approach to the authorship. The chapter commences with notions that the rise of robots and automation lead to the loss of 20 million manufacturing jobs by 2030. As a result, in the researchers' opinion, the education sector faces massive restructuring to manage this predicted visionary future. While examining a happiness scale key, the researchers based in Poland found variance in happiness levels of students affected their capacity to undertake robotics programming. The researchers were keen to further ratify their findings through modelling the Rasch measurement theory (RMT). The data analysis was carried out independently by data analysts, providing a fine-grained interpretation of the correlation between happiness levels and Polish classroom robotics pedagogies.

Social Media Analytics, Learning Analytics and Healthcare Industry: Risky Drinking—Much is written about social media analytics, learning analytics and healthcare industry research projects. However, it is rare to find all three domains in the same study, as each discipline is usually kept separate from another. The concern with this isolation approach is the probability of missing essential synergies, leading to a combined effect more significant than the sum of their particular products. For example, social media analytics may uncover key social media search terms that reflect the language used by groups of people engaging in risky drinking behaviour. At the same time, the view of behaviour through a psychometric testing lens may reveal other things. This chapter describes both social media analytics and learning analytics from the perspective of the healthcare industry. The chapter explains the intersection of social media analytics and learning analytics; in the first instance, the Rasch measurement model is used to discover the probability of agreement relating to human behaviour issues; a case study amplifies health services provision outcomes. The results from the social media analytics revealed that groups' risky drinking behaviour data collected by Talkwalker showed linguistic variations with the double meanings of words used, which affected results. The learning analytics of alcohol consumption included a post hoc comparison using a Bonferroni correction t-test which revealed the mean of never-married people was significantly higher than married people. The significance of these findings demonstrates that data analysis should be open to using a mixture of data analytics tools to reach a finer-grained interpretation to model other complex problems.

The Adoption of the Rasch Measurement Approach as a Learning Analytics Tool in Higher Education—This chapter argues that the Rasch measurement approach has valuable contributions that have not yet been fully used in learning analytics to optimise the learning experience in higher education. The chapter aims to show how the application of the Rasch approach offered some practical solutions for redressing the operational challenges facing the learning analytics field. The results reported from a Ph.D. research project are conducted at four higher education institutions in the Kingdom of Saudi Arabia. Data collected from 162 undergraduates voluntarily participated in a series of 2x3 quasi-experiments. The chapter presents the Rasch-driven analyses and explains its practical implications concerning three key elements in the learning process: learners, context and assessment.

Rasch Model Case Study: Investigating Cognitive Performances of School children's Museum Learning Experiences—This chapter describes how learning analytics was used to improve learning and the environment in which it takes place. There has been much research on measuring, gathering and analysing learner data in formal educational settings. However, its use in informal learning environments like museums appears to be promising. This chapter describes a Ph.D. case study that investigated the cognitive performances of school children aged 10-12 years old during their museum learning experiences by considering the differences in human cognitive preferences. The research study focused on the Dinosaur Walk exhibition in Melbourne Museum as the context of the investigation. The main objective of the research study was to look into the relationship in the form of interactive effect between the students' cognitive preferences and the museum information representation formats in facilitating their museum learning experiences. This research is important as museum learning has been well accepted as informal learning to support formal education. Hence, this chapter serves as an example of managing learning analytics within the informal educational domain. Furthermore, the case study can be exemplary to the education sector and related organisations or businesses towards understanding, planning and assessing students' performances and preparing a better way of facilitating/supporting learners' learning processes and experiences.

Exploring a Framework for Constructivist Design with the eLearning Space—Learning designers and academics require ways to create e-learning courses rapidly. An alternative to time-consuming, content-heavy resources is to design authentic learning experiences that derive from constructivist design. This study aimed to investigate the optimisation of critical learning design elements that informs a constructivist instructional design approach within the Australian e-learning higher education context. A group of subject matter experts (n = 53) from various Australian higher education institutions were purposefully selected to participate in an eDelphi expert panel. In addition, an electronic questionnaire was administered to a larger sample (n = 434). This study employed exploratory sequential design within a mixed-methods approach as narrative and numerical data analysis. The Wilcoxon rank-sum test and Kruskal–Wallis process were applied to validate the perception of respondents identified for the research. The results from the study indicated that learning design elements associated with constructivist concepts such as social learning presence, learning interaction, meta-cognitive load, knowledge-sharing space and knowledge construction might be conducive to the construction of knowledge. The framework presents a modality that can be implemented by instructional designers, academics and online course developers when creating complex e-learning spaces.

Industry 4.0: Learning Analytics Using Artificial Intelligence and Advanced Industry Applications—This chapter explains that learning analytics is a crucial factor in measuring the effectiveness of any educational programme. Industry 4.0, mainly artificial intelligence (AI), applications are integrated into the digital pedagogical techniques and the learning analytics that further enhance and develop the online instructional materials. Therefore, it is crucial to explain the usage of those technologies used for the learning analytics associated with the educational system to show the different uses of AI. This chapter explains the meaning of the learning analytics in Industry 4.0 and Education 4.0 and its affiliated areas, and AI and its related areas with examples from the state-of-the-art corporate applications and solutions based on AI technologies integrated into the learning analytics.

Machine Learning for Learning Analytics for Meta-cognitive Support—The motivation for this chapter is to show the evolutionary nature from earlier student tracking and 'intelligent tutoring systems' until the current stages of advanced learning analytics. This chapter's added value is its illustration of how learning analytics contributes to learners' meta-cognitive awareness and how indirectly a teacher's 'dashboard' allows teachers, parents and remedial teachers to trace the more subtle cognitive entailments in the student's mind. The sources for the chosen direction rest upon earlier research into media support for conceptual learning. This chapter may elicit the discussion on how far technology-driven innovations can bypass the existing repertoire of earlier didactic solutions.

Associate Professor Elspeth McKay
Director, Cogniware
Melbourne, Australia
e.mckay@cogniware.com.au

References

Adams, R. J., & Khoo, S.-T. (1996). *QUEST: The interactive test analysis system* (Vol. 1). Melbourne: Australian Council for Educational Research.

Andrich, D., & Marais, I. (2019). *A course in rasch measurement theory: Measuring in the educational, social and health sciences.* Western Australia: Springer Texts in Education.

Bakkar, M. N. (2016). *An investigation of mobile healthcare (mHealcare) training design for health-care empoyees in Jordan. Ph.D. thesis.* Available from https://researchbank.rmit.edu.au/view/rmit:161776. RMIT University, Australia.

Boone, W. J., Staver, J. R., & Yale, M. S. (2014). *Rasch analysis in the human sciences.* Yale: Springer.

Cohen, J. (1977). *Statistical power analysis for the behavioral sciences* (Revised ed.). New York: Academic Press.

Frisbie, D. A. (1988). Reliability of scores from teacher-made tests. *Instructional topics in educational measurement (ITEMS), spring*, 25–35.

Greller, W., & Drachsler, H. (2012). Translating learning into numbers: Toward a generic framework for learning analytics. *Educational Technology and Society, 15*(3), 42–57.

Griffin, P., & Nix, P. (1991). *Educational Assessment and Reporting: A new approach.* Victoria, Australia: Harcourt Brace.

McKay, E. (2019). *Digital literacy skill development: Prescriptive learning analytics assessment model.* Paper presented at the Research Conference 2019: Preparing students for life in the 21st century: Identifying, developing and assessing what matters, Melbourne Convention and Exhibition Centre, held 4–5 August. https://research.acer.edu.au/research_conference/RC2019/

Meerilees, B., & Miller, D. (1997). The superstore format in Australia: Opportunities and limitations. *Long Range Planning, 306*(6), 899–905.

Popham, W. J. (1978). *Criterion-referenced measurement*. Englewood Cliffs, NJ: Prentice Hall.

Robinson, M., Johnson, A., Walton, D., & MacDermid, J. (2019). A comparison of the polytomous Rasch analysis output of RUMM2030 and R (ltm/eRm/TAM/lordif). *BMC Medical Research Methodology, 19*(1), 36. https://doi.org/10.1186/s12874-019-0680-5

Spector, R. (2005). *Category killers: The retail revolutiokn and its impact on consumer culture.* Boston, Massachusetts: Haravard Press.

Theng, L. F. (2012). *Incorporation of Gagné's instructional design in a student-centred multimedia learning environment.* (Master of Science thesis). Multimedia University, Malaysia. (1585768)

Contents

Editor and Contributors

About the Editor

Elspeth McKay, Ph.D. (https://www.cogniware.com.au) after 25 years at RMIT University, Melbourne, Australia, is moving back into her private consultancy, Cogniware. Her Ph.D. is in computer science and information systems, from Deakin University, Geelong, Australia. She also holds further qualifications in instructional design, computer education and business information systems. She is passionate about designing effective e-learning resources for industry training/reskilling programmes and the education sector. Her project interests involve psychometric testing on the effectiveness of online learning environments. Her work involves developing specialist e-learning tools that implement digital transformation precision. She has published extensively in the research fields of HCI and educational technology, with more than 150 peer-reviewed research papers and several educational technology books, and has been awarded numerous awards for funded projects.
ORCID profile https://orcid.org/0000-0001-7547-9616

Contributors

Ashir Ahmed School of Social Sciences, Media, Film and Education, Melbourne, Australia

Asmidah Alwi School of Creative Industry Management and Performing Arts, Universiti Utara Malaysia, Sintok, Malaysia

Keven Asquith 3D-Geo Pty Ltd, Melbourne, Australia

Mandi Axmann Online Learning Consultant, E-Learning Solutions, Melbourne, Australia

Mahmoud Numan Bakkar Institute of Applied Technology (IAT), Abu Dhabi Vocational Education and Training Institute, Abu Dhabi, UAE

Allaa Barefah Taif University, Taif, Saudi Arabia

Zac Haussegger Swinburne University, Melbourne, Australia

Srećko Joksimović University of South Australia, Adelaide, SA, Australia

Piet Kommers UNESCO Learning Technologies, University of Twente, Enschede, The Netherlands

Tomasz Kopczyński University of Silesia, Katowice, Poland

Djazia Ladjal Practera at Tank Stream Labs, Sydney, NSW, Australia

Ruth Marshall Practera at Tank Stream Labs, Sydney, NSW, Australia

Jenny Martin School of Arts, Federation University Australia, Ballarat, Australia

Elspeth McKay Cogniware, Melbourne, Australia

Abelardo Pardo University of South Australia, Adelaide, SA, Australia

Anna Porczyńska-Ciszewska University of Silesia, Katowice, Poland

Thierry Rakotoarivelo Data61/CSIRO, Eveleigh, NSW, Australia

Claire Scoular Australian Council for Educational Research, Melbourne, Australia

Eugenia Smyrnova-Trybulska University of Silesia, Katowice, Poland

Ian Storey Lecturer Torrens University, Melbourne, Australia

David Towl Access Health and Community, Victoria, Australia

Chen Zhan University of South Australia, Adelaide, SA, Australia

Privacy-Driven Learning Analytics

Srećko Joksimović, Ruth Marshall, Thierry Rakotoarivelo, Djazia Ladjal, Chen Zhan, and Abelardo Pardo

Abstract Data privacy is a central theme in the global dialogue around the application of data science in education. Despite the growing need, research organisations and private companies working on education and learning analytics solutions still rely on ad hoc, red-tape-heavy and inconsistent approaches to privacy protection. This chapter outlines the substantial and growing body of work on data privacy risk measurement and reduction which can help address this problem and enable better use of online learning data with improved privacy risk management. The combination of privacy risk measurement and reduction tools with a sound privacy risk management framework has the potential to manage privacy risk reliably and consistently across all datasets in a pragmatic and cost-effective way as tools evolve to integrate with standard data management infrastructure.

S. Joksimović (✉) · C. Zhan · A. Pardo
University of South Australia, Adelaide, SA, Australia
e-mail: srecko.joksimovic@unisa.edu.au

C. Zhan
e-mail: chen.zhan@unisa.edu.au

A. Pardo
e-mail: abelardo.pardo@unisa.edu.au

R. Marshall · D. Ladjal
Practera at Tank Stream Labs, Sydney, NSW, Australia
e-mail: Djazia.ladjal@practera.com

T. Rakotoarivelo
Data61/CSIRO, Eveleigh, NSW, Australia
e-mail: thierry.rakotoarivelo@data61.csiro.au

E. McKay (ed.), *Manage Your Own Learning Analytics*, Smart Innovation, Systems and Technologies 261, https://doi.org/10.1007/978-3-030-86316-6_1

Introduction

Technology is an integral part of modern education and learning. From the proliferation of technical environments supporting the delivery of blended and fully online courses (Brown et al., 2020; Park & Shea, 2020), the use of academic integrity software, to the broader adoption of artificial intelligence in learning and teaching, technology is deeply interwoven into the fabric of education (de Laat et al., 2020). The impact of the COVID-19 pandemic has highlighted the essential role technology now plays across all levels of education—from university to middle and primary schools. The necessity for educational institutions globally to rapidly pivot to online has resulted in an explosion of learning-related data—both in terms of diversity and volume of learners' digital interactions (Joksimović et al., forthcoming). Emerging data repositories that integrate data from various learning platforms represent a unique opportunity to advance research on education at scale (Gardner et al., 2018). However, a challenge remains with respect to accessing and combining disparate data sources while retaining privacy and ethics. While adherence to privacy and ethical considerations is paramount it impacts the ability to combine heterogeneous data sources and utilise their full potential to further advance our understanding of the learning process in online and digital education models (Joksimović et al., forthcoming).

Data are at the core of learning analytics. As Dawson and colleagues (Dawson et al., 2019) argue, in order for learning analytics to realise its aims and potential, it is critical to depart from small-scale exploratory studies towards a more holistic and integrative systems-level research. For example, to provide a 'truly' personalised learning experience, it is necessary to link diversely structured (e.g. assessment data, demographics) and unstructured (e.g. written artefacts, peer assessment) data sources (Dawson et al., 2019). With recent advances in multi-modal learning analytics (Emerson et al., 2020; Martinez-Maldonado et al., 2020) and in particular with the broader use of sensors to understand learning (Crescenzi-Lanna, 2020), it is possible to collect granular observations of discrete interactions and nuances that are normally overlooked in more traditional learning analytic works. Specifically, novel methods allow for collecting learners' physiological states and behaviours (e.g. heart, brain, skin) (Larmuseau et al., 2020), or capturing social proximity (Yan et al., 2021) and environmental conditions (e.g. location, noise) (Crescenzi-Lanna, 2020). As the data collection processes increase in sophistication there is a parallel risk and degradation of student privacy. The unprecedented opportunities brought about by multimodal learning analytics come with heightened risk and responsibility in ensuring and protecting learner privacy.

Learner privacy and ethical use of data are fundamental to enabling wider adoption of learning analytics in teaching and learning. Specifically, the lack of clear guidelines and transparency concerning ownership and stewardship of learner's data raises serious ethical issues and calls for a unified data privacy framework early in the emergence of the field Ferguson (2012); Siemens (2012). More recently, the European

Union brought into force the General Data Protection Regulation (GDPR). The introduction of the concept of 'privacy by design' by the GDPR represents arguably the most complete regulatory framework in relation to data protection for all European citizens (Regulation, 2016). While existing frameworks noted in learning analytics research (Tsai & Gašević, 2017) have raised the profile and importance of privacy and ethics they are yet to extend beyond a conceptual understanding to more practically oriented solutions. Moreover, despite the increasing focus on privacy and ethics in educational technology, recent reports point to a *"widespread lack of transparency and inconsistent privacy and security practices for products intended for children and students"* (Henein et al., 2020, p. 3). To enable the application of the proposed frameworks in practice and advance educational research at scale, it is crucial to provide methods and processes for implementing privacy-preserving learning analytics (Every Learner Everywhere, 2020).

According to the UN Conference on Trade and Development, 132 countries (66%) worldwide have put in place some form of privacy protection legislation and a further 10% have drafted legislation. Such legislation commonly includes a definition of 'personal' and 'sensitive' information and the broad means by which it should be protected. Australia was one of the earliest countries to introduce data privacy legislation with the 1988 Privacy Act, under which publication or sharing of data is only permissible if personal information, by which an individual can be identified, is removed or obfuscated. Over the past few decades, there has been increasing pressure from grassroots groups such as Privacy International and the Open Rights Group among hundreds of others worldwide to take data privacy and the rights of the digital user more seriously. This has coincided with numerous well publicised data breach scandals resulting in billions of dollars of damages and fines and increasingly broad and onerous privacy legislation such as Europe's General Data Protection Regulation (GDPR) which includes new consideration of downstream recipients of personal data and the results of 'automated decision making' among other things, with breach penalties of up to four per cent of global revenue (Regulation, 2016).

The response to this changing and increasingly impenetrable landscape of risks and legislation has widely varied across organisations in the education and educational technology sectors, depending on risk appetite and access to tools and expertise (Pardo & Siemens, 2014; Tsai & Gašević, 2017). Though there is no lack of information available on data privacy law and associated penalties, there is no common language for measuring and communicating privacy risk or standardised mechanisms for managing it. Without these, data privacy will continue to be a patch-work landscape of ad hoc solutions with variable protection afforded to the end user. Practitioners with higher risk tolerances will take advantage of the new wealth of information becoming available on education platforms to grow their business and increase understanding of the learning process with Learning Analytics, while more risk averse (or often 'risk conscious') organisations fall behind.

There are two possible stages to improving the way in which we address data privacy: Stage 1 is simply to have a standardised way to measure the risk. With a standardised measure of privacy risk, we can remove the variability from our approach to data privacy and communicate as a community about acceptable risk, standards and policies. Stage 2 involves providing measurable, policy driven and provable mitigation mechanisms which are compatible with the techniques of learning analytics while retaining the required utility in the data. In this chapter we will describe this two-stage approach, drawing on well researched and proven methodologies, which will enable a consistent way of measuring and managing privacy risk.

Analytics and Privacy in Educational Research

Data privacy has been identified as one of the main concerns for learning analytics research and practice as well as advancing educational research in general (Ferguson, 2012; Selwyn, 2019). With recent developments in learning analytics and online learning, we are able to collect immense amounts of data about learners and the knowledge processes they engage with (Hillman & Ganesh, 2019). However, this also raises concerns regarding data ownership and privacy.

To date, two broad approaches to addressing issues related to managing student data privacy emerged. Learning analytics and social science researchers in general, primarily focus on developing policies and frameworks for handling student data (Tsai et al., 2020). The difficulties of implementing new and often complex procedures for understanding privacy risk and managing student data are frequently sufficiently onerous to prevent full and robust implementations of these policies and frameworks. Complementary technological solutions are now emerging which offer the opportunity for more comprehensive adoption of policy frameworks with less cost (Gursoy et al., 2017; Kyritsi et al., 2019).

Learning Analytics Data and Associated Risks

The use of data in learning analytics and educational research more broadly came a long way since the early uptake of learning technologies in distance and face-to-face teaching (Joksimović et al., 2019; Siemens, 2012). Specifically, early research in learning analytics and educational data mining was primarily cantered around the utilisation of data collected by various learning management systems (e.g. Bogarín et al., 2014; Hecking et al., 2014; Wolff et al., 2013). Small-scale studies analysing data obtained from a single course, establishing the association between student engagement with learning resources and the final course outcome were common at this early stage (Dawson et al., 2019). In addition, researchers would include data collected by student information systems, such as student demographics, lesson planning, campus attendance or student enrolment data, as well as library attendance (e.g. Brooks & Greer, 2014; Nam et al., 2014). The associated risks and challenges accessing this type of data are further reflected in the lack of scale that was evident in these early works (Dawson et al., 2019).

Alongside the increased utilisation of technology in formal educational settings, informal learning and social media gained significant attention. Although social network analysis has a long tradition in educational research, the scale of networks representing learners' interactions is unprecedented (Joksimović et al., 2019; Papamitsiou et al., 2020). Nationally funded initiatives and projects—such as DiscourseDB (https://discoursedb.github.io/)—emerged with a goal to merge data across various platforms (e.g. blogs, wikis, discussion forums). In addition to understanding who talks to whom, studies focussed on the shared content brought another dimension to enhancing our understanding of learning (e.g. Chiu & Fujita, 2014; Joksimović et al., 2015), as well as increased the associated risks. Thus, from the privacy perspective, multiple challenges emerged. It becomes rather challenging to anonymize the content of the interactions between learners as social media or discussion forum posts could include a reference to another learner, that existing algorithms fail to detect. More importantly, social media tend to blur the line between private and learning/workplace interactions, including more detailed information about personal situations that lay outside of the learning experience.

Further utilisation of various sensors that allow for collection of psychophysiological data, eye gaze or heart rate variability data, to name a few, brought multimodal data to the forefront of learning analytics research (Shankar et al., 2018). This line of research brought novel insights into learners' interactions with their peers and learning environments, that goes beyond more traditional trace data analysis Brishtel et al. (2020), Cornide-Reyes et al. (2020), Sharma et al. (2020). These sensor data bring another level of potential ethical concerns. For instance, a low heart rate variability usually indicates that a person is under higher stress, which could be caused by a learning task at hand (Vogel & Schwabe, 2016). However, existing research further shows that low heart rate variability is also associated with higher incidence of coronary heart disease in the general population or even death in elderly population (Dekker et al., 2000). Often overlooked as well are metadata which can carry a significant amount of contextual information. Risks associated with this kind of data is more complex. For instance, there are no existing privacy-preserving methods that can be used to directly address privacy issues that accompany multimodal data, except for some recent development (Sui & Li, 2017). Such considerations are beyond the scope of this chapter.

Policies and Frameworks

As the use of large-scale data in educational research became more prominent in recent years, there is an associated increase in concerns linked to the privacy of learner data and ethics of their collection for research and practice (Dawson et al., 2019). Early works in privacy and ethics, in educational research and learning analytics, primarily focus on developing policies and frameworks that would allow for exploiting data collected by various learning environments, while still respecting student privacy and building accountability into educational technology (Haythornthwaite, 2017; Hillman & Ganesh, 2019; Reidenberg & Schaub, 2018). Regardless of how the educational landscape evolves, privacy safeguards are necessary to build end-user trust and enable the broader adoption of learning analytics into practice (Reidenberg & Schaub, 2018).

Analysing how existing policies address challenges related to the broader adoption of learning analytics, Tsai and Gašević (2017) provide an overview of eight frameworks related to higher education. In their work, Tsai and Gašević (2017) examine four dimensions associated with learning analytics research including: strategy, obligations, privacy protection and data management and governance and how each dimension is addressed across the existing policies. Specifically, strategy is observed in the context of goal setting in relation to the adoption of learning analytics, methods, evaluation of impact, assurance of validity, communication and support, as well as user roles in the process of adoption. Obligations, on the other hand, represent legal and organisational obligations and privacy protection discusses data anonymity, informed consent and opt-out options (Tsai & Gašević, 2017). Finally, the data management and governance aspect include questions such as data handling process and access to data (Tsai & Gašević, 2017). Without going into detail about other dimensions, data privacy—a primary topic of this chapter—is addressed in each of the studied frameworks. In that regard, data anonymity represents a most commonly addressed issue within the existing frameworks (Drachsler & Greller, 2016; Sclater, 2016; Steiner et al., 2016). Specifically, all the frameworks analysed in Tsai and Gašević (2017), explicitly state that any data being collected will remain anonymous. Some of the frameworks, such as Code of practice for learning analytics developed by JISC (Sclater, 2016), LACE project (Drachsler & Greller, 2016) or Learning Analytics Toolbox (LEA's Box) (Steiner et al., 2016) go beyond this and envision that data must remain anonymous when transferred between various sources or aggregated. Finally, LACE project (Drachsler & Greller, 2016), LEA's Box (Steiner et al., 2016) and Open University UK (Slade, 2016) frameworks further consider the involvement of third parties, saying that "special attention needs to be paid to data anonymity and any data sharing needs to comply with privacy requirements in the current policy as well as institutional and national/international policies" (Tsai & Gašević, 2017, p. 239).

In a more recent study, Brown and Klein (Brown & Klein, 2020), however, show that there is an urgent need for organisations and institutions to *"revise their data governance use policies to represent data systems as they are"* (Brown & Klein, 2020, p. 20). Analysing 151 university policy statements from almost 80 public and private post-secondary institutions in the US, Brown and Klein (Brown & Klein, 2020) reveal several important issues with the existing policies and frameworks, primarily stemming from the bureaucratic nature of those policies. Specifically, the primary goal of these existing frameworks is driven by the goal of protecting institutional liability and reducing potential risk, whereby data are being treated as static artefacts (Brown & Klein, 2020). More importantly, students are being positioned as "informed, agentic partners in their education" (Brown & Klein, 2020, p. 11). It is unclear, as Brown and Klein (Brown & Klein, 2020) further note, how those policies address individual students' needs. Moreover, there is a striking tension between *"how data systems are often conceptualised as static and how they are enacted as dynamic technologies"* (Brown & Klein, 2020, p. 15).

To move towards providing more demonstrable processes, Hoel and colleagues (Hoel et al., 2017) provide an extensive list of requirements that should represent essential components of any learning analytics architecture. Analysing international privacy frameworks, that are currently being turned *"into national laws in various parts of the world"* (Hoel et al., 2017, p. 243), Hoel and colleagues suggest directions for the future discourse around privacy engineering in learning analytics. Likewise, Reidenberg and Schaub (2018) make a series of policy recommendations for building the accountability and oversight, that are 'safeguards for privacy' (Reidenberg & Schaub, 2018, p. 264), into learning technologies. With respect to the technological safeguards, Reidenberg and Schaub (2018) argue that technical mechanisms for ensuring transparent data collection, processing and use, algorithmic accountability and secure handling of student data, are fundamental to data use in education.

Technical Solutions

Several approaches have been proposed recently as feasible solutions to the pressing need for rethinking how learning analytics systems and practices are designed with privacy requirements in mind (Hoel et al., 2017). The existing approaches apply different models for protecting learners privacy, from providing platform-as-a-service architecture (Gardner et al., 2018) and utilising blockchain publicly verifiable system for educational data (Hillman & Ganesh, 2019), to the application of various anonymization and provably private mechanisms (Gursoy et al., 2017). Some examples of the former two approaches are discussed below, while they provide some benefits they do not formally define the offered privacy guarantee. Section "Examples of Re-identification Risk Mitigation Techniques" reviews some of the commonly used anonymization techniques, which have practical limits as their privacy assurance is tied to a specific dataset. Finally, Sect. "Provable Privacy Enhancing Mechanisms" introduces more recent provably private mechanisms, which ground their privacy guarantee in the process used to share/release data, without revealing individual's identity. It is noteworthy that this section discusses non-provable mechanisms. That is, while educational research and learning analytics, in particular, tend to use "anonymised" as a reference to the differentially private, it is important to note that data privacy and confidentiality community clearly define 'anonymisation' as a method that does not guarantee differential privacy (Rocher et al., 2019). The concept of differential privacy will be discussed in Sect. "Provable Privacy Enhancing Mechanisms".

The proliferation of MOOCs and online learning, in general, brought new promises as well as new challenges to educational research. Thus, it is not surprising that one of the first architectures for providing access to privacy-restricted datasets was developed around providing access to MOOC data. Specifically, Gardner and colleagues (2018) introduced MOOC Replication Framework (MORF), a platform-as-a-service architecture for large-scale computational research (e.g. predictive modelling and production rule analysis), while protecting student privacy. In so doing, MORF provides 'execute-against' access to the platform data exports. That is, platform users are able to deploy analysis and run experiments. However, instead of downloading data directly, the platform provides a summary of results. As such, MORF undoubtedly represents a significant step towards the resolution of the existing barriers to enabling replication and scale in educational data research. Nevertheless, being implemented as platform-as-a-service, MORF has a role of an "intermediary" between institutions that generate the data and end users. Architecturally, MORF expects that data would be shared in a raw format, which could have considerable implications in regard to privacy and ethics (Sect. "Policies and Frameworks").

Hillman and Ganesh (2019), on the other hand, proposed a system (Kratos) that provides students and schools an *"immutable log along with comprehensive access to data that is otherwise scattered across systems and vendors"* (Hillman & Ganesh, 2019, p. 5754), placing a particular focus on the aspect of data ownership, data accountability, auditability and transparency. The main rationale behind Kratos is framed around the necessity to integrate data across a wide variety of educational platforms, developed by various vendors, that are being used in K-12 settings. As Hillman and Ganesh (2019) further argue, despite the existence of open educational data standards, there is necessarily no incentive for vendors in K-12 space to adopt any of those. Moreover, data privacy terms and policies are not always clear (Hillman & Ganesh, 2019; Kelly et al., 2019). The solution proposed by Hillman and Ganesh (2019) relies on a defined set of smart contracts, based on the existing agreements between schools and technology vendors. Similar to MORF (Gardner et al., 2018), Kratos also relies on the combination of technical solutions (Interplanetary File System and file decryption) to mitigate issues of data and consequently privacy protection. As such solutions could provide immense benefits to individual users (e.g. students or individual schools), providing insights and control over their data, there is still limited potential for scaling up educational research and making an impact in educational technology research and practice.

Stage 1—Risk Assessment and Privacy

We can all agree that we have a legal and ethical obligation as educators and education technology providers to keep the personally identifiable information (PII) of our students private. The real question is: what does this mean? How far does this obligation go in terms of limiting what we do with our data and what must we do to ensure we are meeting our obligations? Legislation varies considerably, but a safe generalisation of the more comprehensive laws would include a definition of PII as any information by which an individual can be identified. Common sense says that this would include unique identifiers such as driver's license number or email address and specific but not necessarily unique information such as home address and phone number and name. If we remove these obvious identifiers have, we done enough to protect data privacy? Most data practitioners instinctively know the answer is no. With all PII removed there is still a residual and often significant, risk that individuals can be re-identified from the data that remains (Barbaro & Zeller, 2006; Golle, 2006; Narayanan & Shmatikov, 2008).

Many technical mechanisms have been proposed to tackle the challenge of preserving an individual's sensitive information from a dataset while ensuring that such a dataset remains useful for some given data analysis. Some examples of such early techniques include the use of perturbation (e.g. adding some random noise to data points) or generalisation (e.g. merging close data points into bins of fixed ranges or clusters). However, these mechanisms are dependent on the datasets on which they are applied and thus often provide limited privacy assurance and are subject to potential attacks (Machanavajjhala et al., 2007). Simply removing Personal Identifier Information (PII) from a dataset is also not a viable approach to preserve individuals' sensitive information. Indeed, it has been shown in several instances that often simple combinations of available attributes in a dataset may act as quasi-identifier. When coupled with available auxiliary information (e.g. from social media or other publicly known sources) these attribute combinations are sufficient to identify an individual's data record with high certainty.

There are various types of risks that may lead to privacy breaches (e.g. risk of recovering masked attributes, risk of inferring individual membership to a group). This chapter specifically focuses on the risk of re-identifying an individual from a dataset, even in cases when all PII were removed as discussed above. A number of factors will influence such re-identification risk including what the data give away and what other information can be brought into play. Unique data points, unique combinations of attributes, or repeated patterns in series of event data (such as public transport records) can uniquely identify an individual in a de-identified dataset, particularly if you know something else about the individual or can use these patterns to link to another dataset. This was nicely exemplified in 2016 by a study at the University of Melbourne (Culnane et al., 2017), which demonstrated that de-identified Australian public health records could be re-identified in some instances with 'a few mundane facts taken together.' From this dataset of three million people, one researcher was able to identify herself with a high degree of certainty using just her date of birth and those of her two children.

When it comes to re-identification, we seem to suffer from a mass case of 'hindsight vision' whereby every breach appears obvious and easily avoidable after detection. As a result, the *"misconception that this is either a solved problem, or an easy problem to solve"* is still pervasive, particularly in non-government organisations, though the growing number of similar studies is slowly changing opinions. There is also a significant body of research into re-identification of aggregate data, showing that it can also be relatively trivial to do depending on the aggregation levels and the information therein (Gadotti et al., 2019). This is still not as widely understood as it should be, resulting in inadequate uniformly applied threshold rules (such as 'no less than five') and over-confidence in their effect on residual risk Asghar and Kaafar (2020).

Measuring Re-Identification Risk

An important first step to improve our understanding of data privacy risk is to design and use consistent and objective methods to quantify such risk in datasets. Such a privacy risk measure, one that can be applied to any dataset and compared across datasets, will enable us to consistently assess the risk of re-identification so that we can apply appropriate risk reduction mechanisms. We will then be able to use the same uniform measure to verify that the risk has been reduced to an acceptable level based on our risk appetite, potential impact and required use of the data. Beyond this essential basic, a consistent re-identification risk measure will allow us to:

- set context-driven data risk standards and policies applicable to different uses and recipients,
- establish data sharing and exchange policies using a common risk scale,
- automate risk reduction mechanisms in your data management pipelines using standardised privacy risk metadata,
- verify that privacy risk management standards are being met by your partners and suppliers.

Several frameworks proposed methods to qualitatively assess the re-identification risks of a dataset (OKeefe et al., 2017). While these methods provide valuable information to data custodians, they do not offer a consistent and systematic measurement scale that would allow comparisons across datasets or evaluations of simple mitigation measures. On the other hand, other existing contributions offer mathematical definitions of re-identification risk, thus allowing consistent quantification across datasets.

For example, one recent contribution (Masood et al., 2018) proposed a probabilistic method based on Hidden Markov Chain Models (HMM) to quantify the risk of re-identifying individuals from collected Web data. This method takes into account the uniqueness, uniformity and linkability of a user's data. It is generic and can be applied to any event-based datasets (e.g. transaction records, web queries). Other works are using information theory results to design an entropy-based measure of the 'information surprisal' (IS) from a given dataset. Several variants of such IS metrics were proposed and evaluated (Chen et al., 2013). These metrics were further used to design a method to quantify the uniqueness of a user profile on Online Social Networks.

Finally, another contribution further extended the design and use of entropy-based metrics to quantify re-identification risks in many complex data scenarios. These include datasets that provide samples of larger populations, or data perturbed with some random bounded noise, or binning rules. In one specific example, such a metric has been implemented in an operational software, the Re-identifier Risk Ready Reckoner (R4). R4 can operate on unit-record and event-record data and compute scores that quantify the potential risk associated with an attacker using different combinations of background knowledge to attempt the re-identification of an individual. These scores can be used to understand and compare the risk profile of different attributes (and their combinations) within a dataset, across different risk reduction treatments of such dataset and across multiple datasets. R4 provides both a front-end dashboard interface and an Application Programming Interface (API) to support both direct interaction with data custodians and integration within third-party data workflows.

Examples of Re-Identification Risk Mitigation Techniques

Several methods were proposed to mitigate the privacy and confidentiality risks as measured by the metrics discussed above. These methods will modify the original data in the hope of lowering these risks. One of the challenges in designing these methods is to retain enough of the original information in the transformed data to ensure that it is still fit for purpose, e.g. that some given analysis can still be applied to it and return some informative results. This balance between the privacy and the utility of the data is critical to any privacy risk mitigation method and remains an ongoing research question.

Some of the simplest mitigation techniques are based on perturbation. Perturbation mechanisms rely on the addition of random value to an existing data point to provide some level of privacy, i.e. adding 'noise' to the data point. The selection process of that noise to maintain any desired utility in the data is challenging. While these techniques appear to provide some form of privacy, they are susceptible to various statistical attack models, which can easily recover the original data points from the noisy ones (Gadotti et al., 2019).

Generalisation is another family of basic mitigation techniques. Simple examples of generalisation include rounding data points, grouping them into bins of value range, or clustering them and replacing them with their cluster's centroid. More sophisticated generalisation techniques have been recently proposed, such as k-anonymity, l-diversity and t-closeness (Li et al., 2007; Samarati & Sweeney, 1998). While these techniques provide systematic mechanisms to formally define and apply some level of desired privacy to a dataset, several subsequent works discussed successful attack scenarios against them and studied their limits in balancing privacy and utility (Brickell & Shmatikov, 2008).

Stage 2—Towards Provable Privacy

The main issue with many of the earlier privacy risk mitigation techniques, as described in Sect. "Examples of Re-identification Risk Mitigation Techniques", is that their definition of the privacy being provided is specifically tied to the dataset at hand. For example, one limit of such an approach is that it cannot consider all existing or future auxiliary knowledge that an adversary might have and use to circumvent the applied privacy mechanism (Machanavajjhala et al., 2007). Another class of techniques are based on mathematical definitions of privacy guarantees that are independent of the dataset being processed. This de-coupled approach allows these techniques to be resilient against such background knowledge-based attacks. We will further refer to this class of techniques as provably private methods.

Provable Privacy Enhancing Mechanisms

One of the major classes of provable privacy mechanisms is based on the model of Differential Privacy (DP) (Dwork & Roth, 2014). The DP model defines the provided privacy as a property of the mechanism to be applied to a dataset (or a query on a dataset) to allow its release. A DP mechanism applied to a dataset essentially guarantees that its outputs will be very similar with/without the inclusion/removal of any single individual's information in that dataset. Thus, it offers plausible deniability that this individual was (or not) part of the group on which the original data was collected. The earlier works on DP (Dwork & Roth, 2014) provide its mathematical definition along with several proved properties (e.g. composability, post-processing, relaxation), which make it suitable for integration with a wide range of novel provable privacy processes.

One common method to ensure DP is the addition of noise from a specifically designed Laplace distribution to specific parts of the mechanism to be applied to the data. DP mechanisms have been designed for many types of data usages. For example, some works proposed DP-based mechanisms to generate a completely synthetic dataset as a substitute for an original one (Asghar et al., 2020). The synthetic version retains highly accurate levels of information as found in the original one. It could for example preserve how single, pairs or n-tuples of attributes are distributed. Other contributions proposed DP-based mechanisms to allow interactive statistical queries about a dataset to be answered, again with some high accuracy on the results and some interesting properties, such as example supporting streams of data (Perrier et al., 2020). More recent work proposed DP-based additions to existing machine learning algorithms to allow the learning of models with strong privacy assurance. In the context of federated learning, one example proposes a DP-enhanced gradient descent algorithm, which enables the distributed learning of a model from multiple datasets hosted by different data custodians (Wu et al., 2020).

While the theoretical definition and mathematically proved properties of DP provide strong privacy assurance, their implementations into software functions and libraries need to be carefully executed to avoid introducing biases that may weaken such assurance. For example, design and implementations of floating-points in some programming languages may lead to biased distributions of noise over double-precision numbers, which in turn may break DP (Mironov, 2012).

More recent works have started to identify several practical limits to many DP-based mechanisms. For example, (Kifer & Machanavajjhala, 2011) showed that strong correlations in graph datasets can strongly suppress any utility from DP- based mechanisms on such data. This is a significant limitation as real-world graph data (e.g. social networks) often exhibits such strong edge/vertices correlations. Further novel contributions pushed on these limits of DP and proposed new privacy models and definitions, such as the (Pufferfish Privacy Kifer & Machanavajjhala, 2014), the Inferential Privacy (Ghosh & Kleinberg, 2017), or Zero-Knowledge Privacy (Gehrke et al., 2011).

Finally, there are other classes of provable privacy mechanisms that allow dataset release and joint analysis with strong privacy guarantees. Secure Multi-Party Computing techniques are an example of such an alternative class, with for example mechanisms based on homomorphic encryption (Jagannathan & Wright, 2005; Phong et al., 2018), which allow different non-trusting parties to perform joint computation on their collective data without sharing them in a clear format. Privacy-Preserving Record Linkage (PPRL) mechanisms constitute another class of provably private techniques, with for example methods designed around Bloom Filter encoding (Vatsalan et al., 2017).

Privacy-Driven Learning Analytics

Among the most prominent work, Gursoy and colleagues (2017) and Kyritsi and colleagues (2019) rely on differential privacy mechanisms to enable privacy-preserving learning analytics. Both studies focus on two approaches to data privacy: (i) privacy-preserving data publishing and (ii) privacy-preserving data mining. Gursoy and colleagues (2017) experimentally confirmed the obvious trade-offs between privacy and utility. Increasing the level of privacy necessarily leads to reduced accuracy in learning analytics tasks. Moreover, Gursoy and colleagues (2017) showed that anonymization methods - such as k-anonymity (Sweeney, 2002) or l-diversity (Machanavajjhala, et al., 2007)—provide a considerably better balance between utility and privacy, compared to differential privacy-based approach. The challenge of finding the balance between information loss and preserving privacy was further confirmed by Kyritsi and colleagues (2019). In their study, Kyritsi and colleagues (Kyritsi et al., 2019) observed how the utility of the dataset changes after applying k-anonymity and l-diversity data perturbation in the case of simple correlation analysis.

The existing work also confirms the importance of educational theory for learning analytics, as argued in numerous studies before (Gašević et al., 2015; Joksimović et al., 2019; Wise & Shaffer, 2015). Specifically, Kyritsi and colleagues (2019) showed that adding more variables in learning analytics models tends to increase *"the disclosure risk of the whole data"* (ibid., p. 9). The disclosure risk becomes even more prominent in smaller datasets. In that sense, it becomes critical to design our learning analytics around the educational constructs and measures that actually capture learning (Joksimović et al., 2019). Deploying privacy-driven learning analytics and maintaining student privacy would be, in that case, structured around preserving privacy around educational constructs that matter, rather than exposing entire logs with limited value to understanding learning. Among others, Joksimovic and colleagues (Joksimović et al., 2019) posit that the importance of establishing learning analytics research in existing theories stems from the nature of the learning environments predominantly used for learning and capturing learning-related activities. Specifically, the existing platforms *"were primarily designed to support course delivery and not necessarily empowering design for learning"* (Joksimović et al., 2019, p. 52), which results in limited capabilities for designing learning tasks that would elicit various aspects of learning (Carvalho & Goodyear, 2014). In that sense, more data does not necessarily mean better data. Therefore, eliminating the disclosure risk relates to establishing educationally relevant metrics that would allow us to understand learning as a process and as an outcome. Such data would be limited in scope, giving a better chance to develop privacy-driven learning analytics, instead of trying to expose potentially useless and highly revealing data.

Conclusion

There is no technical necessity to implement privacy risk measurement ahead of risk reduction—as solutions they are not necessarily on the same evolutionary path. That said, there are some good pragmatic reasons for doing so. The ability to measure and communicate privacy risk is an important enabler for everyone: beyond the specialist analysts and data scientists, to the daily consumers of data in the form of reports and analyses, those forming and approving policies, or those giving consent to having their data used. It has the potential to provide the vocabulary for intra- and inter-organisational discussions and agreements around data sharing and privacy protection. It can help educate the community about privacy risk and potentially help to counter some of the common over-estimation of the levels of protection the more basic risk reduction approaches afford. This chapter has predominately focussed on the emerging technologies which will make universal, standardised privacy risk management more achievable in the very near future. Alongside these enabling technologies, we will also need to see an evolution of privacy risk awareness and understanding in the communities of people who access, share and consume data.

There is still substantial ground to cover when it comes to finding the ideal "sweet spot" between acceptable levels of privacy risk reduction and utility, or fitness, of the data for its analytical purpose (Gursoy et al., 2017; Kyritsi et al., 2019). Though this is a well-known and heavily researched, issue with differential privacy techniques, it is equally problematic with the other risk reduction approaches discussed here. One potential way forward is to adapt the analytics so that they are optimised for certain risk reduction techniques and can apply risk reduction to their results. A family of learning analytics models which are packaged to address a broad set of questions that are commonly used in the field can be pre-matched with a form of privacy risk reduction which will provide a good level of utility and analysis outcome. For instance, graph data analysis would be matched to risk reduction techniques that are known to work best for this type of data, such as zero-knowledge privacy. There is no silver bullet emerging from the privacy risk reduction research, so a more pragmatic 'best fit' approach will be required if risk reduction techniques are to be widely adopted. For such an approach to have broad appeal and adoption there needs to also be a shift in focus to the specificity of systems engineering, user experience and pragmatic privacy policy development. Privacy-driven learning analytics can help reduce the complexity of a "best fit" approach by incorporating the mechanisms of data privacy risk management thus removing them from the explicit management of the data analyst. Privacy led analytics, coupled with good risk measurement and policy frameworks, can be catalysts for change with respect to our approach to data collection and use (Every Learner Everywhere, 2020).

Data Privacy is a central theme in the global dialogue around ethical Artificial Intelligence and the application of Data Science in Education. Governments around the world are acting on the concerns of their citizens with increasingly onerous and punitive privacy legislation (Borenstein & Howard, 2021; The Institute for Ethical AI in Education, 2021). Despite the growing need, research organisations and private companies working on education and learning analytics solutions still rely on ad hoc, red-tape-heavy and inconsistent approaches to privacy protection. As this chapter outlines, there is a substantial and growing body of work on data privacy risk measurement and reduction which can help address this problem and enable better use of online learning data with improved privacy risk management. The combination of privacy risk measurement and reduction tools with a sound privacy risk management framework has the potential to manage privacy risk reliably and consistently across all datasets in a pragmatic and cost-effective way as tools evolve to integrate with standard data management infrastructure.

References

Asghar, H. J., Ding, M., Rakotoarivelo, T., Mrabet, S., & Kaafar, D. (2020). Differentially private release of datasets using Gaussian Copula. *Journal of Privacy and Confidentiality, 10*(2). https://doi.org/10.29012/jpc.686.

Asghar, H. J., & Kaafar, D. (2020). Averaging attacks on bounded noise-based disclosure control algorithms. *Proceedings on Privacy Enhancing Technologies, 2020*(2), 358–378. https://doi.org/10.2478/popets-2020-0031

Bailey, P., & Sclater, N. (2015). *Code of practice for learning analytics* (p. 6). Jisc. https://www.jisc.ac.uk/guides/code-of-practice-for-learning-analytics.

Barbaro, M., & Zeller, T. (2006). A face is exposed for AOL searcher no. 4417749. *New York Times*. https://www.nytimes.com/2006/08/09/technology/09aol.html.

Bogarín, A., Romero, C., Cerezo, R., & Sánchez-Santillán, M. (2014). Clustering for improving educational process mining. In *Proceedings of the fourth international conference on learning analytics and knowledge* (pp. 11–15). https://doi.org/10.1145/2567574.2567604.

Borenstein, J., & Howard, A. (2021). Emerging challenges in AI and the need for AI ethics education. *AI and Ethics, 1*(1), 61–65.

Brickell, J., & Shmatikov, V. (2008). The cost of privacy: destruction of data-mining utility in anonymized data publishing. In *Proceedings of the 14th ACM SIGKDD international conference on knowledge discovery and data mining* (pp. 70–78). https://doi.org/10.1145/1401890.1401904.

Brishtel, I., Khan, A. A., Schmidt, T., Dingler, T., Ishimaru, S., & Dengel, A. (2020). Mind wandering in a multimodal reading setting: Behavior analysis and automatic detection using eye-tracking and an EDA sensor. *Sensors, 20*(9). https://doi.org/10.3390/s20092546.

Brooks, C., & Greer, J. (2014). Explaining predictive models to learning specialists using personas. In *Proceedings of the fourth international conference on learning analytics and knowledge* (pp. 26–30). https://doi.org/10.1145/2567574.2567612.

Brown, M., & Klein, C. (2020). Whose data? Which rights? Whose power? A policy discourse analysis of student privacy policy documents. *The Journal of Higher Education, 91*(7), 1149–1178. https://doi.org/10.1080/00221546.2020.1770045

Brown, M., McCormack, M., Reeves, J., Brook, D. C., Grajek, S., Alexander, B., Bali, M., Bulger, S., Dark, S., Engelbert, N., Gannon, K., Gauthier, A., Gibson, D., Gibson, R., Lundin, B., Veletsianos, G., & Weber, N. (2020). *2020 Educause horizon report teaching and learning edition*. EDUCAUSE. https://www.learntechlib.org/p/215670.

Carvalho, L., & Goodyear, P. (2014). *The architecture of productive learning networks*. Routledge.

Charles Sturt University. (2015). *CSU learning analytics code of practice*. Charles Sturt University. https://www.csu.edu.au/__data/assets/pdf_file/0007/2160484/2016_CSU_LearningAnalyticsCodePractice.pdf.

Chen, T., Chaabane, A., Tournoux, P. U., Kaafar, M.-A., & Boreli, R. (2013). How much is too much? Leveraging ads audience estimation to evaluate public profile uniqueness. In E. De Cristofaro & M. Wright (Eds.), *Privacy enhancing technologies* (pp. 225–244). Springer.

Chiu, M. M., & Fujita, N. (2014). Statistical discourse analysis of online discussions: Informal cognition, social metacognition and knowledge creation. In *Proceedings of the fourth international conference on learning analytics and knowledge* (pp. 217–225). https://doi.org/10.1145/2567574.2567580.

Cornide-Reyes, H., Riquelme, F., Monsalves, D., Noel, R., Cechinel, C., Villarroel, R., Ponce, F., & Munoz, R. (2020). A multimodal real-time feedback platform based on spoken interactions for remote active learning support. *Sensors, 20*(21). https://doi.org/10.3390/s20216337.

Corrin, L., Kennedy, G., French, S., Buckingham Shum, S., Kitto, K., Pardo, A., West, D., Mirriahi, N., & Colvin, C. (2019). The ethics of learning analytics in Australian higher education. *Accessed Online, 26*. https://researchnow.flinders.edu.au/en/publications/the-ethics-of-learning-analytics-in-australian-higher-education-d.

Crescenzi-Lanna, L. (2020). Multimodal learning analytics research with young children: A systematic review. *British Journal of Educational Technology, 51*(5), 1485–1504. https://doi.org/10.1111/bjet.12959

Culnane, C., Rubinstein, B. I. P., & Teague, V. (2017). *Health data in an open world*. https://arxiv.org/abs/1712.05627.

Dawson, S., Joksimovic, S., Poquet, O., & Siemens, G. (2019). Increasing the impact of learning analytics. In *Proceedings of the 9th international conference on learning analytics and knowledge* (pp. 446–455). https://doi.org/10.1145/3303772.3303784.

de Laat, M., Joksimovic, S., & Ifenthaler, D. (2020). Artificial intelligence, real-time feedback and workplace learning analytics to support in situ complex problem-solving: A commentary. *The International Journal of Information and Learning Technology, 37*(5), 267–277. https://doi.org/10.1108/IJILT-03-2020-0026

Dekker, J. M., Crow, R. S., Folsom, A. R., Hannan, P. J., Liao, D., Swenne, C. A., & Schouten, E. G. (2000). Low heart rate variability in a 2-minute rhythm strip predicts risk of coronary heart disease and mortality from several causes. *Circulation, 102*(11), 1239–1244. https://doi.org/10.1161/01.CIR.102.11.1239

Drachsler, H., & Greller, W. (2016). Privacy and analytics: It's a DELICATE issue a checklist for trusted learning analytics. In *Proceedings of the sixth international conference on learning analytics and knowledge* (pp. 89–98). https://doi.org/10.1145/2883851.2883893.

Dwork, C., & Roth, A. (2014). The algorithmic foundations of differential privacy. *Foundations and Trends in Theoretical Computer Science, 9*(3–4), 211–407. https://doi.org/10.1561/0400000042

Emerson, A., Cloude, E. B., Azevedo, R., & Lester, J. (2020). Multimodal learning analytics for game-based learning. *British Journal of Educational Technology, 51*(5), 1505–1526. https://doi.org/10.1111/bjet.12992

Every Learner Everywhere. (2020). *Guiding principles and startegies for learning analytics implementation.* Tyton Partners. https://www.everylearnereverywhere.org/resources/learning-analytics-strategy-toolkit/.

Ferguson, R. (2012). Learning analytics: Drivers, developments and challenges. *International Journal of Technology Enhanced Learning, 4*(5–6), 304–317. https://doi.org/10.1504/IJ$_{TEL}$.2012.051816

Ferguson, R. (2019). Ethical challenges for learning analytics. *Journal of Learning Analytics, 6*(3), 25–30. https://doi.org/10.18608/jla.2019.63.5.

Gadotti, A., Houssiau, F., Rocher, L., Livshits, B., & Montjoye, Y. -A. de. (2019a). When the signal is in the noise: Exploiting Diffix's sticky noise. *28th USENIX Security Symposium (USENIX Security 19)*, 1081–1098. https://www.usenix.org/conference/usenixsecurity19/presentation/gadotti.

Gardner, J., Brooks, C., Andres, J. M., & Baker, R. (2018). Replicating MOOC predictive models at scale. *Proceedings of the Fifth Annual ACM Conference on Learning at Scale.* https://doi.org/10.1145/3231644.3231656

Gašević, D., Dawson, S., & Siemens, G. (2015). Let's not forget: Learning analytics are about learning. *TechTrends, 59*(1), 64–71.

Gehrke, J., Lui, E., & Pass, R. (2011). Towards privacy for social networks: A zero-knowledge based definition of privacy. *Theory of Cryptography Conference*, 432–449.

Ghosh, A., & Kleinberg, R. (2017). Inferential privacy guarantees for differentially private mechanisms. *Innovations in Theoretical Computer Science Conference, 9*(1–9), 3.

Golle, P. (2006). Revisiting the uniqueness of simple demographics in the US population. In *Proceedings of the 5th ACM workshop on privacy in electronic society* (pp. 77–80). https://doi.org/10.1145/1179601.1179615.

Gürses, S. (2014). Can you engineer privacy? *Communications of the ACM, 57*(8), 20–23. https://doi.org/10.1145/2633029

Gursoy, M. E., Inan, A., Nergiz, M. E., & Saygin, Y. (2017). Privacy-preserving learning analytics: Challenges and techniques. *IEEE Transactions on Learning Technologies, 10*(1), 68–81. https://doi.org/10.1109/TLT.2016.2607747

Haythornthwaite, C. (2017). An Information policy perspective on learning analytics. In *Proceedings of the seventh international learning analytics and knowledge conference* (pp. 253–256). https://doi.org/10.1145/3027385.3027389.

Henein, N., Willemsen, B., & Woo, B. (2020). *The state of privacy and personal data protection, 2020–2022.* Gartner Report. https://www.gartner.com/en/documents/3989495/the-state-of-privacy-and-personal-data-protection-2020-2.

Hillman, V., & Ganesh, V. (2019). Kratos: A secure, authenticated and publicly verifiable system for educational data using the blockchain. *IEEE International Conference on Big Data (big Data), 2019*, 5754–5762. https://doi.org/10.1109/BigData47090.2019.9006190

Hoel, T., Griffiths, D., & Chen, W. (2017). The influence of data protection and privacy frameworks on the design of learning analytics systems. In *Proceedings of the seventh international learning analytics and knowledge conference* (pp. 243–252). https://doi.org/10.1145/3027385.3027414.

Jagannathan, G., & Wright, R. N. (2005). *Privacy-preserving distributed k-means clustering over arbitrarily partitioned data.* (pp. 593–599). https://doi.org/10.1145/1081870.1081942.

Joksimović, S., Kovanović, V., & Dawson, S. (2019). The journey of learning analytics. *HERDSA Review of Higher Education, 6*, 27–63.

Joksimović, S., Kovanović, V., Jovanović, J., Zouaq, A., Gašević, D., & Hatala, M. (2015). What Do CMOOC participants talk about in social media? A topic analysis of discourse in a CMOOC. In *Proceedings of the fifth international conference on learning analytics and knowledge*, 156–165. https://doi.org/10.1145/2723576.2723609

Joksimović, S., Siemens, G., Coyle, D., De Laat, M., Dawson, S., Richey, M., Kovanović, V., Pardo, A., & Alexei, F. (2021). Understanding knowledge processes: Data for artificial intelligence research (DAIR) infrastructure. *IEEE Transactions on Learning Technologies*, 14.

Kelly, G., Graham, J., Bronfman, J., & Garton, S. (2019). *2019 State of edtech privacy report*. Common Sense Media. https://privacy.commonsense.org/resource/2019-state-of-edtech-privacy-report.

Kifer, D., & Machanavajjhala, A. (2014). Pufferfish: A framework for mathematical privacy definitions. *ACM Trans. Database Syst.*, *39*(1), 3:1–3:36.

Kifer, D., & Machanavajjhala, A. (2011). No free lunch in data privacy. *ACM SIGMOD International Conference on Management of Data*, 193–204.

Kyritsi, K. H., Zorkadis, V., Stavropoulos, E. C., & Verykios, V. S. (2019). The pursuit of patterns in educational data mining as a threat to student privacy. *Journal of Interactive Media in Education, 2019*(1).

Larmuseau, C., Cornelis, J., Lancieri, L., Desmet, P., & Depaepe, F. (2020). Multimodal learning analytics to investigate cognitive load during online problem solving. *British Journal of Educational Technology, 51*(5), 1548–1562. https://doi.org/10.1111/bjet.12958

Li, N., Li, T., & Venkatasubramanian, S. (2007). t-Closeness: Privacy Beyond k-Anonymity and l-Diversity. In *2007 IEEE 23rd international conference on data engineering*, (pp. 106–115). https://doi.org/10.1109/ICDE.2007.367856.

Machanavajjhala, A., Kifer, D., Gehrke, J., & Venkitasubramaniam, M. (2007). L-Diversity: Privacy beyond k-anonymity. *ACM Transactions on Knowledge Discovery from Data, 1*(1), 24–24. https://doi.org/10.1145/1217299.1217302

Martinez-Maldonado, R., Echeverria, V., Fernandez Nieto, G., & Buckingham Shum, S. (2020). From data to insights: A layered storytelling approach for multimodal learning analytics. In *Proceedings of the 2020 CHI conference on human factors in computing systems*, (pp. 1–15). https://doi.org/10.1145/3313831.3376148.

Masood, R., Vatsalan, D., Ikram, M., & Kaafar, M. A. (2018). Incognito: A method for obfuscating web data. In *Proceedings of the 2018 world wide web conference* (pp. 267–276). https://doi.org/10.1145/3178876.3186093.

Mironov, I. (2012). On significance of the least significant bits for differential privacy. In *Proceedings of the 2012 ACM conference on computer and communications security* (pp. 650–661). https://doi.org/10.1145/2382196.2382264.

Nam, S., Lonn, S., Brown, T., Davis, C. -S., & Koch, D. (2014). Customized course advising: Investigating engineering student success with incoming profiles and patterns of concurrent course enrollment. In *Proceedings of the fourth international conference on learning analytics and knowledge* (pp. 16–25). https://doi.org/10.1145/2567574.2567589

Narayanan, A., & Shmatikov, V. (2008). Robust de-anonymization of large sparse datasets. In *2008 IEEE symposium on security and privacy (SP 2008)*, (pp. 111–125). https://doi.org/10.1109/SP.2008.33.

OKeefe, C. M., Otorepec, S., Elliot, M., Mackey, E., & OHara, K. (2017). The de-identification decision-making framework. *Technical Reports EP173122 and EP175702*. https://publications.csiro.au/rpr/download?pid=csiro:EP173122&dsid=DS3.

Papamitsiou, Z., Giannakos, M. N., & Ochoa, X. (2020). From childhood to maturity: Are we there yet? Mapping the intellectual progress in learning analytics during the past decade. In *Proceedings of the tenth international conference on learning analytics and knowledge* (pp. 559–568). https://doi.org/10.1145/3375462.3375519.

Pardo, A., & Siemens, G. (2014). Ethical and privacy principles for learning analytics. *British Journal of Educational Technology, 45*(3), 438–450. https://doi.org/10.1111/bjet.12152

Park, H., & Shea, P. (2020). A ten-year review of online learning research through Co-citation analysis. *Online Learning, 24*(2). https://doi.org/10.24059/olj.v24i2.2001.

Patterns in a blended learning course. In *Proceedings of the fourth international conference on learning analytics and knowledge* (pp. 173–182). https://doi.org/10.1145/2567574.2567584.

Perrier, V., Asghar, H. J., & Kaafar, D. (2020). Private continual release of real-valued data streams. In *Proceedings of the 2020 network and distributed system security symposium (NDSS)*. https://www.ndss-symposium.org/ndss-paper/private-continual-release-of-real-valued-data-streams/.

Phong, L. T., Aono, Y., Hayashi, T., Wang, L., & Moriai, S. (2018). Privacy-preserving deep learning via additively homomorphic encryption. *IEEE Transactions on Information Forensics and Security, 13*(5), 1333–1345.

Regulation, G. D. P. (2016). Regulation EU 2016/679 of the European parliament and of the council of 27 April 2016. *Official journal of the European Union*. Available at: http://Ec.Europa.Eu/Justice/Data-Protection/Reform/Files/Regulation_oj_en.Pdf (Accessed September 20, 2017).

Reidenberg, J. R., & Schaub, F. (2018). Achieving big data privacy in education. *Theory and Research in Education, 16*(3), 263–279. https://doi.org/10.1177/1477878518805308

Rocher, L., Hendrickx, J. M., & de Montjoye, Y.-A. (2019). Estimating the success of re-identifications in incomplete datasets using generative models. *Nature Communications, 10*(1), 3069. https://doi.org/10.1038/s41467-019-10933-3

Samarati, P., & Sweeney, L. (1998). *Protecting privacy when disclosing information: K-Anonymity and Its enforcement through generalization and suppression*. Harvard Data Privacy Lab.

Sclater, N. (2016). Developing a code of practice for learning analytics. *Journal of Learning Analytics, 3*(1), 16–42. https://doi.org/10.18608/jla.2016.31.3.

Selwyn, N. (2019). What's the problem with learning analytics? *Journal of Learning Analytics, 6*(3), 11–19. https://doi.org/10.18608/jla.2019.63.3.

Shankar, S. K., Prieto, L. P., Rodríguez-Triana, M. J., & Ruiz-Calleja, A. (2018). A review of multimodal learning analytics architectures. In *2018 IEEE 18th international conference on advanced learning technologies (ICALT)* (pp. 212–214). https://doi.org/10.1109/ICALT.2018.00057.

Sharma, K., Leftheriotis, I., & Giannakos, M. (2020). Utilizing interactive surfaces to enhance learning, collaboration and engagement: Insights from learners' gaze and speech. *Sensors, 20*(7). https://doi.org/10.3390/s20071964.

Siemens, G. (2012). Learning analytics: Envisioning a research discipline and a domain of practice. In *ACM international conference proceeding series* (pp. 4–8). https://doi.org/10.1145/2330601.2330605.

Slade, S. (2016). *The open university ethical use of student data for learning analytics policy* (p. 11). https://doi.org/10.13140/RG.2.1.1317.4164.

Slade, S., & Prinsloo, P. (2013). Learning analytics: Ethical issues and dilemmas. *American Behavioral Scientist, 57*(10), 1510–1529. https://doi.org/10.1177/0002764213479366

Solove, D. J. (2005). A taxonomy of privacy. *U. Pa. l. Rev., 154*, 477.

Steiner, C., Kickmeier, M., & Albert, D. (2016). LEA in private: A privacy and data protection framework for a learning analytics toolbox. *Journal of Learning Analytics*, 66–90. https://doi.org/10.18608/jla.2016.31.5.

Sui, P., & Li, X. (2017). A privacy-preserving approach for multimodal transaction data integrated analysis. *Neurocomputing, 253*, 56–64. https://doi.org/10.1016/j.neucom.2016.09.130

Sweeney, L. (2002). k-anonymity: A model for protecting privacy. *International Journal of Uncertainty, Fuzziness and Knowledge-Based Systems, 10*(05), 557–570.

The Institute for Ethical AI in Education. (2021). *The ethical framework for AI in education* (p. 15). The Institute for Ethical AI in Education. https://www.buckingham.ac.uk/research-the-institute-for-ethical-ai-in-education/.

Tsai, Y. -S., & Gašević, D. (2017). Learning analytics in higher education—Challenges and policies: A review of eight learning analytics policies. In *Proceedings of the seventh international learning analytics & knowledge conference* (pp. 233–242). https://doi.org/10.1145/3027385.3027400.

Tsai, Y. -S., Whitelock-Wainwright, A., & Gašević, D. (2020). The privacy paradox and its implications for learning analytics. In *Proceedings of the tenth international conference on learning analytics & knowledge* (pp. 230–239). https://doi.org/10.1145/3375462.3375536.

Vatsalan, D., Sehili, Z., Christen, P., & Rahm, E. (2017). Privacy-preserving record linkage for big data: Current approaches and research challenges. In A. Y. Zomaya & S. Sakr (Eds.), *Handbook of big data technologies* (pp. 851–895). Springer International Publishing. https://doi.org/10.1007/978-3-319-49340-4_25.

Vogel, S., & Schwabe, L. (2016). Learning and memory under stress: Implications for the classroom. *Npj Science of Learning, 1*(1), 1–10.

Wise, A. F., & Shaffer, D. W. (2015). Why theory matters more than ever in the age of big data. *Journal of Learning Analytics, 2*(2), 5–13. https://doi.org/10.18608/jla.2015.22.2.

Wolff, A., Zdrahal, Z., Nikolov, A., & Pantucek, M. (2013). Improving retention: Predicting at-risk students by analysing clicking behaviour in a virtual learning environment. In *Proceedings of the third international conference on learning analytics and knowledge* (pp. 145–149). https://doi.org/10.1145/2460296.2460324.

Wu, N., Farokhi, F., Smith, D., & Kaafar, M. A. (2020). The value of collaboration in convex machine learning with differential privacy. *IEEE Symposium on Security and Privacy (SP), 2020*, 304–317.

Yan, L., Martinez-Maldonado, R., Cordoba, B. G., Deppeler, J., Corrigan, D., Nieto, G. F., & Gasevic, D. (2021). Footprints at school: Modelling in-class social dynamics from students' physical positioning traces. In *LAK21: 11th international learning analytics and knowledge conference* (pp. 43–54). Association for Computing Machinery. https://doi.org/10.1145/3448139.3448144.

Srećko Joksimović is a Senior Lecturer in Data Science at the Education Futures, University of South Australia. His research is centred around augmenting the abilities of individuals to solve complex problems in collaborative settings. Srećko is particularly interested in evaluating the influence of contextual, social, cognitive and affective factors on groups and individuals as they solve complex real-world problems. In so doing, he utilises a wide range of methods from machine learning, artificial intelligence and natural language processing, as well as data science and social computing in general.

ORCID profile https://orcid.org/0000-0001-6999-3547

Ruth Marshall is Director of R&D and Data Integrity at Australian Education Technology company Practera where she works with universities and research organisations to innovate new ways of delivering high quality work-integrated learning and lifelong learning skills development at scale. Ruth started her career with a UK A.I. startup Cognitive Applications. She has since worked in the startup community in the UK, US and Australia involved in technical and executive roles with 11 + startups. Prior to her current role, she spent five years as a commercialisation specialist with CSIRO/Data61. She has also held R&D, management consulting and CIO positions for large international corporations including British Telecom, Westpac, Accenture and Novartis Pharmaceuticals. She has a Bachelors degree in Computing and Artificial Intelligence from the University of Sussex, England.

ORCID profile https://orcid.org/0000-0002-4040-3360

Thierry Rakotoarivelo is a senior research scientist at CSIRO Data61. His research focuses on data privacy and confidentiality. He is interested in the design and use of frameworks for privacy risk assessment, the development of provable privacy mechanisms and the study of utility/privacy trade-off in specific application domains. He has been leading projects focussing on data privacy in the Government, EdTech and Energy sectors, where he designed and delivered novel technology in areas such as privacy and confidentiality risk quantification or provably private synthetic data generation. He previously worked on network protocols, mission critical wireless networks and frameworks for reproducible experiments in computer science. Thierry received his Ph.D. in computer science from the University of New South Wales in cotutelle with the Institut National Polytechnique of Toulouse (INPT). Prior to that, he was a senior research engineer at the Motorola Australian Research Centre (MARC), where he developed several patents in the domain of mission critical wireless networks.

Djazia Ladjal has a Ph.D. in Astrophysics with 10 + years of experience in research and data modelling. She has also worked in Media and Marketing, developing machine learning algorithms for various commercial enterprises before pivoting to the EdTech space and Learning Analytics. For the past year, she has been leading the data science research at the EdTech company Practera. This exposure to academia and industry provides her with a unique perspective on the intersection between fundamental research on applied technological solutions. She is passionate about data and how to apply complex models to solve everyday problems.

ORCID profile https://orcid.org/0000-0003-4906-231X

Chen Zhan is a research associate and data scientist at the Centre for Change and Complexity in Learning (C3L) at the University of South Australia (UniSA). He obtained his master degree in software engineering from the University of Science and Technology of China (USTC) in 2016 and completed his Ph.D. in computer and information science from UniSA in 2020. His research centres around data mining and its application in various disciplines, including environmental science, economics, bioinformatics and pharmacoepidemiology. Recently, he focuses on artificial intelligence and learning analytics for the education sector. In particular, he contributes to the research in the development of products and services of privacy-preserving learning analytics for the education technology industry.

ORCID profile https://orcid.org/0000-0002-4794-8339

Abelardo Pardo is Professor and Dean of Programs (Engineering), UniSA STEM at the University of South Australia. His research interests include the design and deployment of technology to increase understanding and improve digital learning experiences. More specifically, his work examines the areas of learning analytics, personalised active learning and technology for student support. ORCID profile https://orcid.org/0000-0002-6857-0582.

ORCID profile https://orcid.org/0000-0002-6857-0582

Using Log Stream Data and Item Response Theory to Understand Proficiency in Collaboration

Claire Scoular

Abstract The nature of skills such as collaboration is complex, given that there are internal processes at play and inferences need to be made to interpret explicit behaviours observed from intentionally designed assessment tasks. The theoretical concept of the paths presented in this chapter is to a priori define the processes being sought in the process stream data and interpret their substantive meaning in relation to the skill being measured. This chapter centres on a learning analytical approach to reduce complex skills into simpler but observable components that can be represented as aspects of the target construct and scaled to understand the growth of such skills. The method outlined in this chapter presents an approach that transforms process stream data of collaboration into formative assessment data that can be analysed using Item Response Theory. The application of this model in turn allows the interpretation of the data as levels of proficiency that we can use to map or monitor progress in collaborative skills.

Assessing 21st Century Skills

There is increasing awareness that twenty-first century skills are important for success in education and work environments (O'Neil et al., 2004). With this demand, there is a greater need for the measurement of these skills and traditional methods may be insufficient. Measuring less explored socio-cognitive skills requires innovative approaches, blending different disciplines such as computer science, learning analytics and psychometrics to effectively assess these skills (Scoular & Care, 2017).

One particular twenty-first century skill of focus in this chapter is collaboration. There is increasing demand to work well with others and to work globally (O'Neil et al., 2004). Consequently, collaboration skills that allow effective working in groups have been identified as increasingly important for success in education and work environments (Singh-Gupta & Troutt-Ervin, 1996). Collaboration has been shown

C. Scoular (✉)
Australian Council for Educational Research, Melbourne, Australia
e-mail: Claire.scoular@acer.org

E. McKay (ed.), *Manage Your Own Learning Analytics*, Smart Innovation, Systems and Technologies 261, https://doi.org/10.1007/978-3-030-86316-6_2

to enhance cognitive development (Webb, 1989; Zhang, 1998) and has been demonstrated to have advantages in encouraging accountability, ability to ask questions and justify responses, flexibility in problem solving and reflective skills (Baghaei et al., 2007; Soller, 2001; Webb et al., 1998).

Social interaction or awareness during cognitive tasks such as problem solving has been considered beneficial for some time. Several prominent researchers highlighted the learning benefits to the individual of interaction with other humans, suggesting that placing learners in a social context is a core strategy for developing complex cognitive skills such as problem-solving competency (Glaser, 1992; Vygotsky, 1986; Wittrock, 1989). When learners work collaboratively to solve problems, they think through the problem and the processes more explicitly during their interaction with others, which leads to a greater conceptual understanding and leads them to manage tasks more effectively (Darling-Hammond, 2003).

There is research to suggest that learners process information differently when they work in groups compared to working independently (King et al., 1997). Social interactions make learners' understanding explicit and learners usually improve their comprehension through discussion with others, elaborating and negotiating with others to reach shared understanding (Van Boxtel et al., 2000). Collaborative tasks such as asking questions, peer mentoring and providing feedback can help learners to solve problems or finish tasks they may have otherwise not been able to solve or complete and therefore allow them to move towards higher levels of proficiency (King et al., 1997). Social interactions while working through complex tasks can provide additional ideas and shared meaning that an individual would not achieve without communicating with others (OECD, 2013).

One challenge in measuring skills such as collaboration is that they are complex, particularly given there are internal processes at play and inferences need to be made to interpret explicit behaviours observed from intentionally designed assessment tasks. Data captured from online activities of individuals are used for many functions and now are a familiar feature of computer use. Only recently has this facility been generated at scale and is now providing new insights and perspectives in measurement. What is referred to as "process stream data" captures actions in online activity and retains a timestamped record (Adams et al., 2015). In understanding skills, process stream data provides innovative ways to capture, store and analyse on-task activities and behaviours so that inferences can be drawn about ability within and across twenty-first century skills (Alom et al., 2017).

Quantitative metrics such as count of actions and measures of time are useful, although they do not provide the full picture of the nature of what is being measured. Quantitative analysis of log files often informs very little regarding the quality of the interactions. A single-events approach is simple both from a data-processing and measurement modelling perspective (see Adams et al., 2015). However, using single events as indicators of behaviour loses substantial amounts of information because sequential order is not captured. For example, when assessing the quality of collaborative interaction, a sequence of actions and chat, such as the placement of the chat between significant actions, could be of more value than the frequency of chat. There is evidence from the literature that information gained from sequences of actions can provide important insights into underlying complex behaviours (Cooper & Shallice, 2006; Kuniyoshi et al., 1994; Lohman & Ippel, 1993; Yamauchi & Beer, 1994). The events and sequence of interactions in the data can provide detailed indicators of the dynamics between participants. Specific combinations of behaviours identified from the analytics will be presented in this chapter to demonstrate that they can indicate the process in which people complete tasks and solve problems. Not only do sequences yield richer information, they allow for capturing complex constructs that are procedural in nature.

Beyond identifying the importance of certain skills, there is little guidance on where, when and how to develop, teach and train these skills. This is partly due to a lack of understanding of the nature of skills, including how they develop and change over time and which aspects are fundamental building blocks. Education systems frequently emphasize summative assessments that centre on overall score points. These are valuable but typically do not contribute to individual learning or development of growth in skills. However, identifying specifically what a learner needs to know or be able to do can better inform how instructional support can be adapted to support people to progress towards their goals. In other words, the focus should be on learning progressions, rather than on the scores.

Method

Overview

The method outlined in this chapter presents an approach that transforms process stream data of collaboration into formative assessment data that can be analysed using Item Response Theory. The application of this model in turn allows the interpretation of the data as levels of proficiency that can be used to map or monitor progress in collaborative skills.

Develop Tasks that Capture Log Stream Data

The definition of a skill has implications for assessment task design and good measurement practice indicates that a clear definition of the construct the assessment is measuring should be identified before task design (Scoular et al., 2017). This chapter highlights an assessment that measures collaboration in problem solving contexts in 11–16 year old learners working in groups of two (see Scoular, 2017). The assessment is designed on the understanding of the ACER definition of collaboration (see Fig. 1) in which it is identified as “Collaboration refers to the capacity of an individual to contribute effectively in a group” (p. 12, Scoular et al., 2020). Within the collaboration definition, there is a specification of three strands: building shared

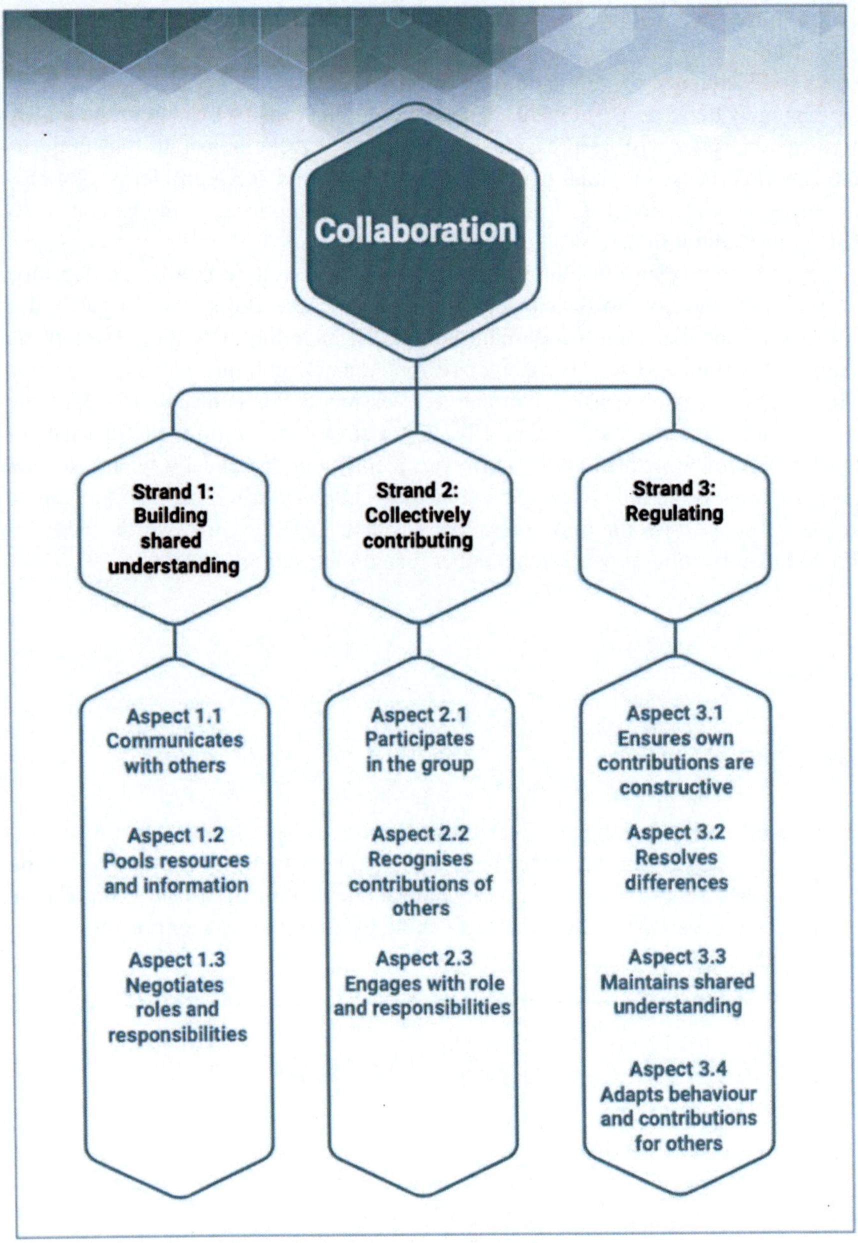

Fig. 1 Collaboration definitional framework from Scoular et al. (2020)

understanding, collectively contributing and regulating and within each strand, there are associated aspects. The focus of the assessment is developmental, equipping learners to draw on and demonstrate collaborative skills in ambiguous problem-solving contexts and where neither the most efficient paths nor final solutions are known. The focus, therefore, is not solution-focussed, but skills-focussed.

Figure 2 presents an excerpt from the assessment task. In this task, learners have to navigate around a map to deliver boxes on a scooter to local facilities such as the post office and bank. This task presents asymmetrical perspectives of the problem, where one learner can view a map and the other learner can view the controls for the scooter. The learner with the controls does not know how the controls work (i.e. which button moves the scooter in which direction) and together they have to explore and work out the function of the controls through delivery of the boxes. The learner viewing the map can see the scooter moving and can direct but cannot see which controls move in which direction.

This task can elicit many collaborative behaviours linked to the aspects outlined in the Scoular et al. (2020) framework. Communication in this task is critical, particularly while building a shared understanding of the problem and will give an indication of how proficient learners are. It is also important that learners realise

Page 1: Explore the problem space			
Instructions	You need to deliver the boxes to the required locations. Work with your partner to determine the controls and deliver all of the boxes.		
Resources	View of map including three destinations and scooter (*student A*)	Four navigation controls (*student B*)	View of list of destinations Chat box *(both students)*
Actions		Press four navigation controls (*student B*)	Chat *(both students)*
Example of indicators and associated subskills	Interaction – Initiates communication Collects elements of information – Actively seeks information Transactive memory – Communicating shared information		
	Page 1 Student A View		**Page 1 Student B View**

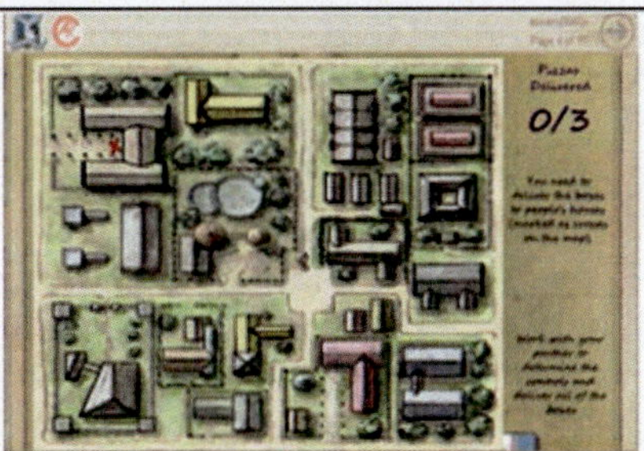

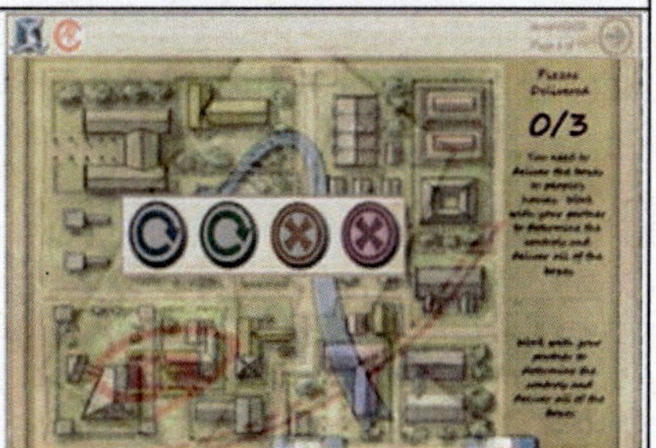

Fig. 2 Excerpt from collaboration task (from Scoular, 2017)

early on that they have very different perspectives and resources. The way in which learners manage their collective resources will give an indication of their collaborative proficiency. Ideally, learners will ask one another questions about their respective resources, seeking further information about how to pool the information. Learners may be observed providing their partners with key information, such as which direction to move, followed by their partner asking them a question and the learner providing additional information. When key information is shared in such a discursive way, we can infer that learners are thinking as a group and depending on one another for transfer of information and knowledge.

Turning Log Stream Data into Measurement Items

All of the actions and chat events that occur in the assessment are captured in log stream data in real time and timestamped. This allows algorithms to be created to search through the log stream data and identify combinations of those actions and chats that link to the definitional framework and can be interpreted as demonstrating specific collaborative aspects. For example, one behaviour pattern is an action-chat-action event sequence in which both learners identify different answers in relation to the principle governing the problem and after chat events, one learner ultimately changed their answer to match the same as the other learner. It is inferred from this pattern of behaviour that learners have realised their answers are different and they seek to resolve the differences; they chat about their respective answers until they come to a consensus and one learner changes their answer to match their partner's. As such, this indicator has been mapped to the resolving differences aspect (aspect 3.2) in the regulating strand of the collaboration framework (Scoular et al., 2020). Figure 3 presents excerpts from the log stream data in which we can observe this sequence of action-chat-action behaviour patterns relating to resolving differences. Excerpt 1 shows that learner B had entered '50' into their answer and learner A has entered '100' into theirs (in brackets). The arrow indicates that later, on the same page and after at least one chat event from each learner, learner B changes their answer to '100', matching learner A's answer. In Excerpt 2 we can observe a similar action-chat-action sequence of events but this time from another set of learners. Learner A and learner B have different answers, '15' and '10' respectively. There is the presence of at least one chat event each before learner B changes their answer to '15' (indicated by the arrow). Excerpt 3 presents another example, again with a different pair of learners. Learner A has entered '15' and learner B has entered '16'; there are chat events from each before one learner changes their answer (learner B changes to '15').

Algorithms were developed in order to capture these behaviour patterns in the log stream data. For this particular pattern example, the algorithm first searches for the presence of rule confirm for the learner and for their partner before comparing them. If both are present and the answers are different then the search continues, if this condition is absent then the search discontinues and is recorded as not applicable.

excerpt 1

team_id	student_id	task_id	role	page	event	arg1	arg2	arg3	outcome	timestamp
auvelr0025	auvelr0025a	BALANCE	A	2	summary	0	3	104		23/06/2015 11:09
auvelr0025	auvelr0025b	BALANCE	B	2	rule-confirm	50	0	0		23/06/2015 11:09
auvelr0025	auvelr0025a	BALANCE	A	2	rule-confirm	100				23/06/2015 11:09
auvelr0025	auvelr0025a	BALANCE	A	2	chat	wat did you put for mass... 100?				23/06/2015 11:09
auvelr0025	auvelr0025b	BALANCE	B	2	chat	I say 50				23/06/2015 11:09
auvelr0025	auvelr0025b	BALANCE	B	2	chat	AND YOU ARE ACCIDENT PRONE :)				23/06/2015 11:09
auvelr0025	auvelr0025a	BALANCE	A	2	chat	I MEAN I Think it is 100, because if you add to position 3...				23/06/2015 11:10
auvelr0025	auvelr0025a	BALANCE	A	2	chat	hahahaha doh				23/06/2015 11:10
auvelr0025	auvelr0025a	BALANCE	A	2	chat	yea look at the formula!!				23/06/2015 11:10
auvelr0025	auvelr0025b	BALANCE	B	2	rule-confirm	ok got it!				23/06/2015 11:11
auvelr0025	auvelr0025b	BALANCE	B	2	rule-confirm	100	0	0		23/06/2015 11:11
auvelr0025	auvelr0025a	BALANCE	A	2	rule-confirm	100				23/06/2015 11:11

Excerpt 2

team_id	student_id	task_id	role	page	event	arg1	arg2	arg3	outcome	timestamp
auxmir0024	auxmir0024b	WAREHOUSE	B	5	rule-confirm	10	0	0		1/08/2015 14:33
auxmir0024	auxmir0024a	WAREHOUSE	A	5	rule-confirm	15	0	0		1/08/2015 14:33
auxmir0024	auxmir0024a	WAREHOUSE	A	5	chat	I put 15!	0	0		1/08/2015 14:33
auxmir0024	auxmir0024b	WAREHOUSE	B	5	chat	no! I said 10	0	0		1/08/2015 14:33
auxmir0024	auxmir0024b	WAREHOUSE	B	5	chat	2 times two is 4				1/08/2015 14:34
auxmir0024	auxmir0024b	WAREHOUSE	B	5	chat	that times 2 i 8				1/08/2015 14:34
auxmir0024	auxmir0024b	WAREHOUSE	B	5	chat	minus 1 is 7				1/08/2015 14:34
auxmir0024	auxmir0024a	WAREHOUSE	A	5	chat	2x2x2x2-1?				1/08/2015 14:34
auxmir0024	auxmir0024b	WAREHOUSE	B	5	chat	yeah?				1/08/2015 14:34
auxmir0024	auxmir0024a	WAREHOUSE	A	5	chat	15?				1/08/2015 14:34
auxmir0024	auxmir0024b	WAREHOUSE	B	5	chat	only three 2's				1/08/2015 14:34
auxmir0024	auxmir0024a	WAREHOUSE	A	5	chat	four boxes (2x2x2x2)				1/08/2015 14:34
auxmir0024	auxmir0024a	WAREHOUSE	A	5	chat	three boxes 2x2x2				1/08/2015 14:35
auxmir0024	auxmir0024b	WAREHOUSE	B	5	chat	Oooooh				1/08/2015 14:35
auxmir0024	auxmir0024a	WAREHOUSE	A	5	chat	2x2x2x2-1				1/08/2015 14:35
auxmir0024	auxmir0024b	WAREHOUSE	B	5	chat	yup				1/08/2015 14:35
auxmir0024	auxmir0024a	WAREHOUSE	A	5	chat	16-1=15				1/08/2015 14:36
auxmir0024	auxmir0024b	WAREHOUSE	B	5	chat	so 15?				1/08/2015 14:36
auxmir0024	auxmir0024a	WAREHOUSE	A	5	chat	yeah				1/08/2015 14:36
auxmir0024	auxmir0024b	WAREHOUSE	B	5	chat	Cool!				1/08/2015 14:37
auxmir0024	auxmir0024a	WAREHOUSE	A	5	rule-confirm	15	0	1		1/08/2015 14:37
auxmir0024	auxmir0024b	WAREHOUSE	B	5	rule-confirm	15	0	1		1/08/2015 14:37

Excerpt 3

team_id	student_id	task_id	role	page	event	arg1	arg2	arg3	outcome	timestamp
auvelr0025	auvelr0025a	WAREHOUSE	A	2	chat	I got 15 for last box, you?				1/08/2015 14:36
auvelr0025	auvelr0025b	WAREHOUSE	B	2	rule-confirm	16				1/08/2015 14:36
auvelr0025	auvelr0025a	WAREHOUSE	A	2	rule-confirm	15				1/08/2015 14:36
auvelr0025	auvelr0025b	WAREHOUSE	B	2	chat	16!!??				1/08/2015 14:36
auvelr0025	auvelr0025b	WAREHOUSE	B	2	chat	No				1/08/2015 14:36
auvelr0025	auvelr0025a	WAREHOUSE	A	2	chat	It got 15 again. Trying to work out why...				1/08/2015 14:36
auvelr0025	auvelr0025b	WAREHOUSE	B	2	chat	15??	0	1		1/08/2015 14:36
auvelr0025	auvelr0025a	WAREHOUSE	A	2	chat	I think it is my superior skills				1/08/2015 14:36
auvelr0025	auvelr0025b	WAREHOUSE	B	2	chat	Can't work out formula as I'm too tired. 15 okay with				1/08/2015 14:36
auvelr0025	auvelr0025a	WAREHOUSE	A	2	chat	aok				1/08/2015 14:36
auvelr0025	auvelr0025b	WAREHOUSE	B	2	rule-confirm	15	0	1		1/08/2015 14:36
auvelr0025	auvelr0025b	WAREHOUSE	B	2	chat	Cool. Done!				1/08/2015 14:36
auvelr0025	auvelr0025a	WAREHOUSE	A	2	request-page	P153				1/08/2015 14:36

Fig. 3 Excerpt from log stream data presenting behaviour pattern examples

The algorithm then searches for the second condition, a chat event from the learner after condition one. Then the algorithm searches for the third condition, a rule that confirm that matches the partners. In short, the search algorithm is presented as:

Condition1: Find different rule confirms for SELF and PARTNER

Condition2: Find chat event from SELF after condition1 on same page

Condition3: Find final rule confirms for SELF and PARTNER and both have same value

If all conditions are present then 1; else 0

If no condition1 then not applicable (-1).

Each of these algorithms capturing behaviour patterns is mapped to aspects in the collaboration framework. Table 1 presents examples of behaviour patterns and their respective mapping to the framework. Once each algorithm has searched the log stream data it will record output for each. The output of one of three values: 0, 1 or −1. By default, each algorithm will be recorded as −1 which represents that item as 'missing'. If an algorithm is not applicable to that task or learner it will be recorded as −1 (missing). Reasons for non-applicability include, for example, that the learner did not begin the page in that task it was relevant to or that the search algorithm was

Table 1 Examples of patterns and respective mapping to the framework

Strand from framework	Aspect from framework	Behaviour patterns	Inference
1. Building shared understanding	1.2. Pools resources and information	Presence of pass resource	Shares own resources with others
1. Building shared understanding	1.1. Communicates with others	Chat from self first, followed by chat from partner	Initiates communication with others
1. Building shared understanding	1.1. Communicates with others	Presence of response to question from partner	Responds to questions from others
2. Collectively contributing	2.2. Recognises contributions of others	Partner communicates key information and learner incorporates that information into their answer	Incorporates contributions of others
3. Regulating	3.2. Resolves differences	Different answer from partner, chat from both learners, then one changes answer	Resolves conflicts
3. Regulating	3.2. Resolves differences	Different answer from partner, chat from self, no change in either answer	Attempted but failed negotiation

discontinued partway through as the first step criteria were not present. If the search algorithm identifies an item for that learner is present, then a '1' is recorded for that item against that task for that learner. If it is absent then a '0' is recorded. Recording a '1' or '0' indicates the presence or absence of that item in the log stream data respectively. A record of '1' for an item suggests that the learner possesses some of the construct and specifically some ability in relation to the subskill to which that indicator is mapped to.

The process outlined in this method results in a record table for each learner indicating whether each behaviour pattern was present or absent for them. Through the algorithms, each pattern is essentially turned into an item, much like that on a traditional test. It is from here that we can analyse these items using Item Response Theory (IRT) to test the quality of the items in our test, how reliable they are—how accurately they are at capturing the skill of collaboration- and how valid they are—how strong the inferences about the aspects being measured are.

Results

This chapter explores whether the items work together to form a single variable that represents the construct of collaboration. To do this a model from the Item Response Theory (IRT) family of models was used, the Rasch model. The Rasch model is a commonly used one-parameter model for dichotomously scored items that was introduced by Danish mathematician Georg Rasch (1960). His model is based on the assumptions of a unidimensional construct and existence of increasing function that describes the relationship between item difficulty, represented by the single parameter δ and learner ability represented by the single parameter θ. Rasch presents a model of the probability of item response success based on the learner's ability estimate (Wu & Adams, 2006). The model suggests that a learner with greater ability should have greater probability of solving an item than a learner with lower ability (Rasch, 1960). This model presents both items and learners on the same scale, or continuum, with the positioning of item difficulty location on that scale defined by the probability that a learner answers the item correctly is at 0.5 (Wu & Adams, 2006). Learners with higher ability will have greater probability of answering an item correctly. The item difficulty and person ability estimates are mapped onto a single interval scale, both measured using the same units, called logits. The Rasch model assumes only one parameter for the item (difficulty) and one parameter for the person (ability). Therefore, it is assumed that a learner's ability to perform an indicative behaviour is affected only by their ability on the construct and not by any other factors such as motivation or personal attributes unrelated to the construct under assessment.

Items can be investigated for consistency within the model to support the construct validity of the test. If the items do not present in a cohesive manner they may need to be adjusted or omitted to allow for clearer interpretation of the construct. Items omitted may have perfectly acceptable statistical qualities but if they deviate from the dimension under measurement then they may not be representative of the construct in question and are, therefore, not relevant to the current assessment.

Fit of Items to the Rasch Model

The first purpose in applying the Rasch model to this data is to identify whether the items are of good quality and are consistent with the requirements of the measurement model. That is, whether when learner ability increases the probability of the learner demonstrating more sophisticated collaborative skills also increases. To determine this, the weighted mean square residual goodness of fit (weighted MNSQ) for each item is reviewed (see Wright & Stone, 1979). The weighted MNSQ is used as Wright and Masters (1982) identified the unweighted values as sensitive to unexpected responses from learners who found the item either too easy or too difficult. The weighted MNSQ are identified as effective measures to determine the compatibility of the model with the data. They are based upon residuals describing the deviance between the observed probability of a correct (present) response from the expected response as predicted by the model. To derive the fit statistics, the average of the sum of squared differences between observed and expected scores is weighted by the variance. This measure of fit indicates whether or not the items are assessing a unidimensional underlying construct. Therefore, in this chapter, the fit indicates whether or not the items in the assessment are each measuring learners' collaboration ability. The expected value of the weighted MNSQ is 1.0. The accepted fit range for this chapter is 0.7 to 1.3 taken from Adams and Khoo (1993). There were 75 items in the assessment outlined in this chapter and according to this criterion all of the items in the assessment 'fitted' with the Rasch measurement model (see Table 2 for a sample of items). The good item fit indicated that the probability of a learner obtaining a correct (present) response for that item, increases with the proficiency of collaboration of that learner.

Table 2 Summary of descriptive statistics and calibration output for a sample of items from the collaboration assessment

Item number	Percentage correct (%)	Fit (weighted MNSQ)	Item difficulty
1	54	0.95	−0.23
2	65	1.00	−0.77
3	54	0.96	−0.23
4	61	1.04	−0.55
5	68	0.98	−0.89
6	61	0.91	−0.55
7	71	0.78	−1.09
8	52	1.13	−0.13
9	68	1.05	−0.93
10	62	0.86	−0.59
11	87	0.93	−2.20
12	87	0.93	−2.20
13	49	1.01	0.01
14	17	1.15	1.77
15	23	0.94	1.30
16	29	1.04	0.96
17	39	1.08	0.50
18	54	0.97	−0.23
19	87	0.78	−2.20
20	69	1.04	−0.97
21	49	1.11	−0.01
22	41	1.09	0.36
23	70	0.96	−1.01
24	41	0.98	0.39
25	50	0.97	−0.02
26	60	0.99	−0.51
27	35	1.06	0.69
28	40	1.06	0.43

A sample of item statistics is provided in the interest of brevity and space. Full details can be provided upon request.

Wright Map

It is also important to check if the items cover the range of learner ability. This can be achieved by examination of the Wright map (Wilson & Draney, 2002). The Wright map places the item and learner estimates onto a single scale, using logits as the scaling unit (an arbitrary unit used to enable the location of the two variables on the same metric). It presents the items in increasing order of sophistication in relation to learner ability and can be viewed as two vertical histograms. It is also referred to as a variable map and displays the spread of items.

The Wright map for the assessment outlined in this chapter is presented in Fig. 4. The left side of the figure displays the distribution of learner ability as a histogram ranging from −2.26 to 1.78. The right side of the figure displays the distribution of the item difficulties ranging from −2.20 to 2.83. Items 32, 38 and 50 are at the top of the map indicating they are the most difficult. These items extend beyond

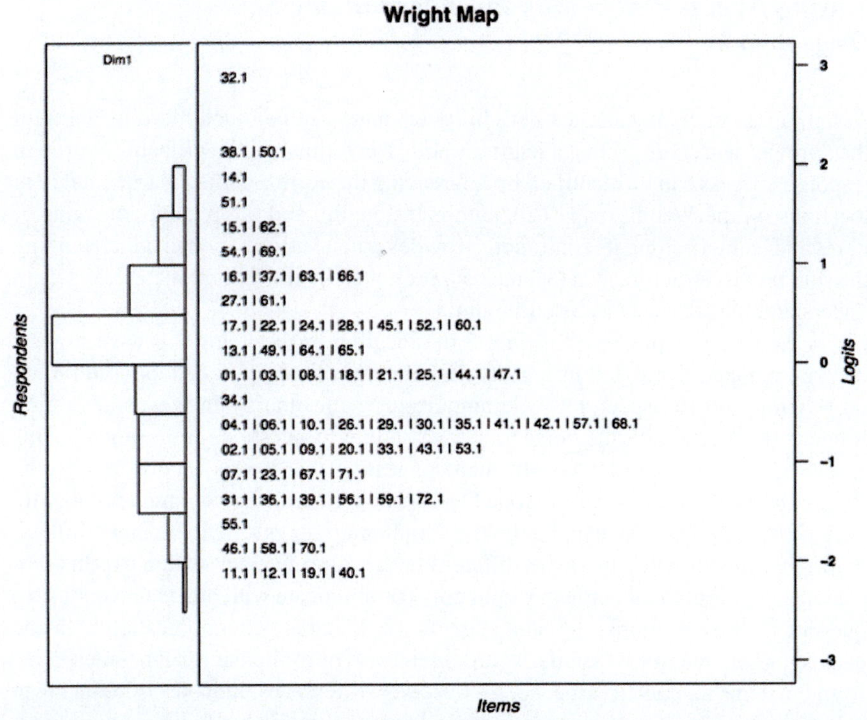

Fig. 4 Wright map for the collaboration assessment

the distribution of the learner ability estimates suggesting these three items were too difficult for the learners. Items 11, 12, 19 and 40 are at the bottom of the map indicating they were the easiest items. The learner ability distribution extends lower than these items so it is difficult to discriminate between those learners at this very low level, although there are only a few of them.

In order to interpret the scale, the item parameters are constrained with the mean item difficulty set at zero. The mean learner ability estimate is −0.055, indicating that on average the learner ability level was lower than the overall assessment difficulty. However, this difference is minimal. The variance of learner ability estimates is 0.71 and the variance of item difficulties is 1.26. The lower mean learner ability and lower variance of learner ability estimates suggest that the overall test was slightly difficult for learners. Given that the learners undertaking the assessment would not have had prior experience with the type of problems or nature of this assessment before this is to be expected. In general, the items are well distributed in relation to the learner ability estimates suggesting a reasonably well targeted assessment.

Interpreting the Assessment Data

Moving from a Wright Map to Understanding Skill Development

Assigning an interval-scale learner ability estimate can be much more meaningful than providing a score. When a learner's ability is estimated, the probability of item response success can be identified by referencing the corresponding learner and item locations on the Wright map. This unidimensionality and the relative locations of items and learners along its continuum provides crucial information in understanding the skill and its structure. That is, the data presented by the Wright map allows us to understand the construct as a continuum.

For example, as presented in Fig. 5 the location of the items and what aspects they are mapped to can tell us a great deal about proficiency in collaboration. Item 70.1 is mapped to aspect 1.1—Communicates with others—in the collaboration framework, specifically the behaviour 'responding to questions from others'. This item is quite far down the Wright map suggesting that it was a fairly easy item. By comparison, another item mapped to aspect 1.1 'initiates communication with others' (item 37.1) is much higher up the Wright map suggesting it is a more difficult item. This suggests that there are different levels of proficiency within aspects. For example, less proficient learners would only communicate with others in response to questions, whereas, more proficient learners would initiate communication with and between group members. Similarly, different levels of proficiency can be interpreted from the items mapped to aspect 3.2—Resolves differences. Item 4.1 is in the lower half of the Wright map and is measuring learners attempting but failing negotiations. By comparison, item 14.1 is one of the most difficult items at the top of the Wright map

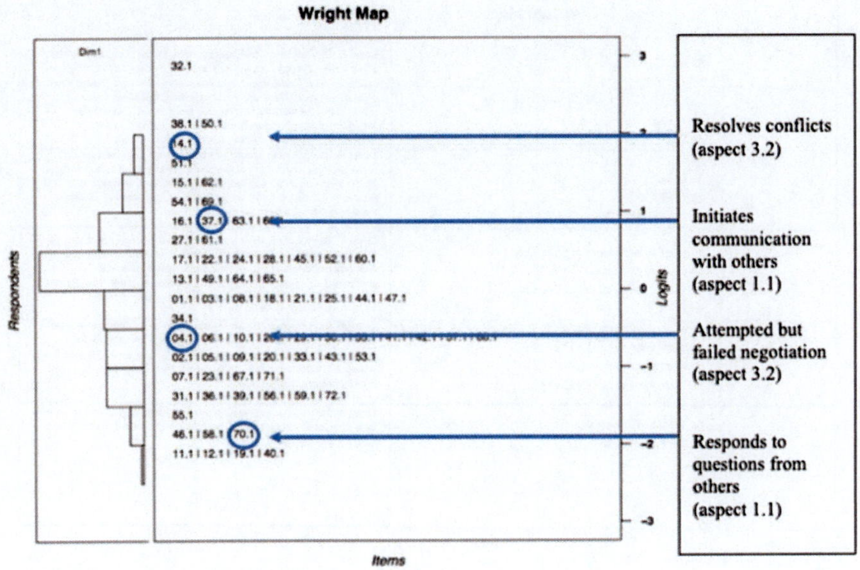

Fig. 5 Interpreting specific items' difficulty in the Wright map

and is measuring 'resolves conflicts'. These item locations suggest that less proficient learners attempt to negotiate with others but often cannot resolve differences, in contrast to high proficiency learners who can resolve differences.

The Wright Map presented in this chapter allows interpretation of the ordering of the items as a progression of collaborative skill. The positioning of items on that continuum can be interpreted to better understand the construct at different levels of proficiency. This allows meaningful information to be relayed back to educators and learners, in relation to what they can and cannot do given a relative estimation. As the items in this chapter relate to behaviours, it can provide educators with real-time data regarding the social and cognitive behaviours their learners are demonstrating. This level of information extends beyond that gathered and interpreted through static tasks or multiple-choice tests. If the expectation of educators is to teach skills such as collaboration, then detailed data about what their learners can demonstrate is imperative and can complement the evidence already being gathered. Having knowledge of the behaviours expected at higher levels of ability can help educators to develop and teach interventions that are targeted to learner needs.

It is also possible to interpret the Wright map as levels of proficiency by interpreting groups of items where they appeared to cluster at similar levels on the scale. Figure 6 identifies cut off points on the scale where natural breaks appear between clusters of items that have similar item difficulty levels. These lines indicate where a discernible change in item difficulty was associated with a change in the kind of skill (or ability) required to demonstrate the associated behaviour pattern. These cut scores

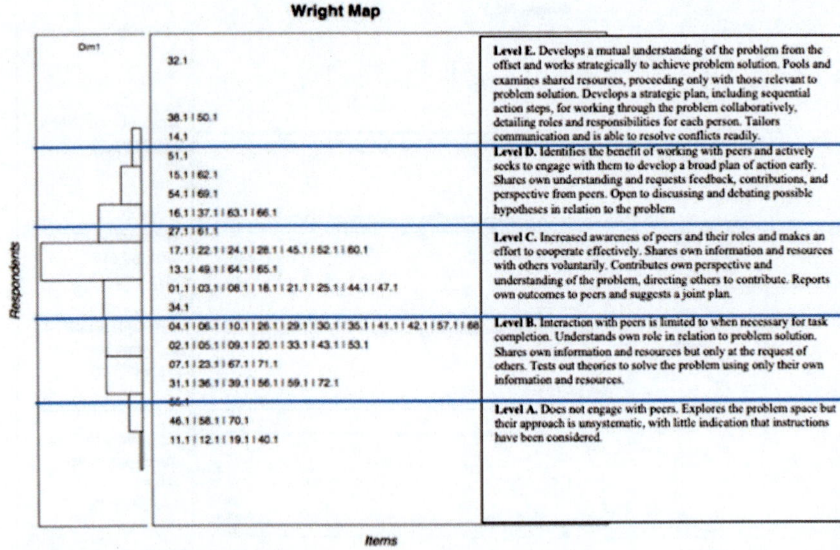

Fig. 6 Interpreting proficiency from the Wright map

and clusters of indicators enable an interpretation of the scale as a developmental progression (Griffin, 2007).

Interpretation of the items as per the aspects of skill involved in the item clusters is elaborated in the text descriptions in Fig. 6. The descriptions can be interpreted as a level of progress in collaboration from novices (A) to experts (E). For example, novices would work independently, unaware of the benefit of engaging with peers and thus would not be likely to solve a collaborative problem. As learners increase in proficiency, they engage more with others and gradually realize the benefit of collaboration. Expert collaborative problem solvers depend upon their peers, develop a mutual understanding of the problem and work strategically to solve it. Negotiation is a critical component of collaboration, but only proficient learners can harness conflicts towards a positive outcome.

A Broader Sense of the Skill

It is important to note that the skill above is interpreted through the lens of a single assessment. It is likely that not every item in a cluster will tell the same conceptual story as the others. While the item may fit statistically, it can be omitted from this part of the interpretation on conceptual grounds. Similarly, no single assessment is expected to measure the entirety of a construct. For example, the assessment presented in this chapter did not measure the aspects of 'adapts behaviours and contributions for others' or 'negotiates roles and responsibilities'. Therefore, interpreting a single assessment in terms of proficiency levels will likely present gaps in the representation of the skill as a whole.

A broader sense of a skill can be achieved through the development of learning progressions. Learning progression as is used to describe how growth in a particular area can be demonstrated and how learners move from early, to more advanced application and understandings. Learning progressions are focused on assessing and monitoring learner growth over time and are underpinned by an understanding that learners of the same age can be a very different points in their learning and development. Learning progressions are not the same as or extensions of learning standards; rather, they focus explicitly on the building blocks learners need to master before they can achieve complex skill forms. When assessments provide information about where learners are in their understanding at the time of assessment, they also provide a basis for monitoring individual progress over time.

To be clear, learning progressions do not describe a single pathway to achieving proficiency in a skill, but instead, provide a picture of the typical path learners follows in developing particular skills. A developmental approach to learning assumes that there is a typical (but not inevitable) pathway that describes and maps the progress of a learner through stages of increasing knowledge, skills and understanding. The aim of the approach is to move learning forward along this path. Figure 7 presents an excerpt of a learning progression for collaboration that has been developed using data from multiple assessments, including the assessment outlined in this chapter (Scoular et al., 2020).

Conclusion

Measuring complex twenty-first century skills such as collaboration can be a challenging endeavour requiring innovative approaches. The approach outlined in this chapter demonstrates a methodological process to identify behaviour patterns from log stream data using search algorithms and interpret this data using Item Response Theory to enable inferences to be made about levels of proficiency in collaboration. As stated in this chapter, no one assessment can paint the larger picture, but each individual assessment of collaboration can provide a piece of the puzzle in understanding this complex skill. The work ahead is iterative. Our understanding of twenty-first century skills needs to be documented so that tasks can be designed appropriately to match this understanding. The more robust assessments of twenty-first century skills that are developed, the more we can corroborate, validate and evolve that understanding. If we are to continue to value these skills in the twenty-first century and beyond, efforts in understanding how these skills develop and what different levels of proficiency look like will need to be applied and the utilisation of process stream data and Item Response Theory is a valuable place to start.

Skill level	Building shared understanding	Collectively contributing	Regulating
Medium	Learners ask for justification of responses or perspective provided. (Aspect 1.1)	Learners acknowledge that others may have a different perspective, and that based on these perspectives, others' contributions may be beneficial to the group as a whole. They understand and incorporate the contributions of others into their own work. (Aspect 2.2)	Learners identify own strengths and weaknesses in relation to the progress of the group task as whole. (Aspect 3.1) Learners make constructive but unsuccessful attempts to resolve differences. (Aspect 3.2) Learners act to maintain shared understanding such as by reiterating or finalising goals, strategy, and roles in more complex tasks. (Aspect 3.3) Learners require feedback from others or explicit requests before they modify or tailor their communication style or behaviour. (Aspect 3.4)
Low–Mid	Learners ask questions or for clarification from others. They will communicate about the related task and respond to contributions of others. (Aspect 1.1) Learners identify that they may not have all of the information required and pool some resources and information with others. (Aspect 1.2) Learners negotiate roles but without considering the expertise, information, or skills held by other group members. (Aspect 1.3)	Learners participate in all necessary tasks throughout the task. Learners maintain a single strategy throughout. Learners collaborate successfully to achieve a straightforward goal. (Aspect 2.1) Learners understand that others may have an alternative perspective. They listen to and acknowledge the perspective of others. (Aspect 2.2) Learners show a willingness and readiness to be involved in the group. They take responsibility for some of the actions determined by their role and provide feedback on their individual tasks. (Aspect 2.3)	Learners reflect on the quality and relevance of their own contributions. (Aspect 3.1) Learners discuss differences of opinion or perspective with others and give careful consideration of the views of others. They comment on differences, but are often unable to resolve them. (Aspect 3.2) Learners act to maintain shared understanding through reiterating goals, strategy, and roles in basic tasks. (Aspect 3.3)

Fig. 7 Excerpt from collaboration learning progression (from Scoular et al., 2020)

References

Adams, R. J., & Khoo, S. T. (1993). *Quest: The interactive test analysis system*. Melbourne, Australia: Australian Centre for Educational Research.

Adams, R. J., Vista, A., Scoular, C., Awwal, N., Griffin, P., & Care, E. (2015). Automatic Coding Procedures. In P. Griffin & E. Care (Eds.), *Assessment and teaching of 21st century skills: Methods & approach* (Vol. 2, pp. 115–132). Springer.

Alom, B. M. M., Scoular, C., & Awwal, N. (2017). Generic log files and algorithms developed for educational multiplayer games. *International Journal of Advanced Computer Research, 7*(33), 171–193.

Baghaei, N., Mitrovic, A., & Irwin, W. (2007). Supporting collaborative learning and problem solving in a constraint-based CSCL environment for UML class diagrams. *International Journal on Computer-Supported Collaborative Learning, 2*(2), 159–190.

Cooper, R. P., & Shallice, T. (2006). Hierarchical schemas and goals in the control of sequential behavior. *Psychological Review, 113*(4), 887–916.

Darling-Hammond, L. (2003). *The learning classroom: Theory into practice*. School of Education, Standford University. Retrieved from https://www.learner.org/courses/learningclassroom/support_pages/index.html

Glaser, B. (1992). *Basics of grounded theory analysis*. Sociology Press.

Griffin, P. (2007). The comfort of competence and the uncertainty of assessment. *Studies in Educational Evaluation, 33*, 87–99.

King, F. J., Rohani, F., & Goodson, L. (1997). *Statewide assessment of listening and verbal communication skills, information literacy skills, and problem-solving skills.* Florida State University.

Kuniyoshi, Y., Inaba, M., & Inoue, H. (1994). Learning by watching: Extracting reusable task knowledge from visual observation of human performance. *IEEE Transactions on Robotics and Automation, 10*(6), 799–822.

Lohman, D. F., & Ippel, M. J. (1993). Cognitive diagnosis: From statistically based assessment toward theory-based assessment. In J. R. Frederiksen, R. J. Mislevy, & I. I. Bejar (Eds.), *Test theory for a new generation of tests* (pp. 41–71). Lawrence Erlbaum Associates.

O'Neil, H. F., Chuang, S., & Chung, G. K. W. K. (2004). Issues in the computer- based assessment of collaborative problem solving. *Assessment in Education, 10*, 361–373.

Organisation for Economic Co-operation and Development (OECD) (2013). *PISA 2015: Draft collaborative problem solving framework.* Retrieved from http://www.oecd.org/callsfortenders/Annex%20ID_PISA%202015%20Collaborative%20Problem%20Solving%20Framework%20.pdf.

Rasch, G. (1960). *Probabilistic models for some intelligence and attainment tests.* The University of Chicago Press.

Scoular, C. & Care, E. (2017). Teaching of twenty-first century skills: Implications at system levels in Australia. *Assessment and teaching of 21st century skills: Research and applications*, volume 3 (pp 32–46). Dordrecht: Springer Science and Business Media.

Scoular, C. (2017). *From log file analysis to item response theory: An assessment template for measuring collaborative problem solving.* Ph.D. Thesis, University of Melbourne. https://minerva-access.unimelb.edu.au/handle/11343/13006.

Scoular, C., Duckworth, D., Heard, J., & Ramalingam, D. (2020). *Collaboration: Skill development framework.* Australian Council for Educational Research. https://research.acer.edu.au/ar_misc/42.

Scoular, C., Care, E., & Hesse, F. (2017). Designs for operationalizing collaborative problem solving for automated assessment. *Journal of Educational Measurement, 54*(1), 12–35.

Singh-Gupta, V., & Troutt-Ervin, E. (1996). Preparing students for teamwork through collaborative writing and peer review techniques. *Teaching English in the Two Year College, 23*, 127–136.

Soller, A. (2001). Supporting social interaction in an intelligent collaborative learning system. *International Journal of Artificial Intelligence in Education, 12*, 40–62.

Van Boxtel, C., Van der Linden, J., & Kanselaar, G. (2000). Collaborative learning tasks and the elaboration of conceptual knowledge. *Learning and Instruction, 10*(4), 311–330.

Vygotsky, L. S. (1986). *Thought and language.* The MIT Press.

Webb, N. M., Nemer, K. M., Chizhik, A. W., & Sugrue, B. (1998). Equity issues in collaborative group assessment: Group composition and performance. *American Educational Research Journal, 35*(4), 607–651.

Webb, N. M. (1989). Peer interaction and learning in small groups. *International Journal of Education Research, 13*, 21–39.

Wilson, M., & Draney, K. (2002). A technique for setting standards and maintaining them over time. In S. Nishisato, Y. Baba, H. Bozdogan, & K. Kanefugi (Eds.), *Measurement and multivariate analysis* (pp. 325–332). Springer-Verlag.

Wittrock, M. C. (1989). Generative processes of comprehension. *Educational Psychologist, 24*(4), 345–376.

Wright, B. D., & Stone, M. H. (1979). *Best test design.* Chicago, USA: MESA Press.

Wright, B. D., & Masters, G. N. (1982). *Rating scale analysis: Rasch measurement*. Chicago, USA: MESA Press.

Wu, M. L., & Adams, R. J. (2006). Modelling mathematics problem solving item responses using a multidimensional IRT model. *Mathematics Education Research Journal, 18*(2), 93–113.

Yamauchi, B. M., & Beer, R. D. (1994). Sequential Behavior and Learning in evolved dynamical neural networks. *Adaptive Behavior, 2*(3), 219–246.

Zhang, J. (1998). A distributed representation approach to group problem solving. *Journal of American Society of Information Science, 49*(9), 801–809.

Claire Scoular's work focuses on the improvement of assessment practices through the application of psychological measurement. Specifically, her expertise revolves around the assessment and teaching of twenty-first century skills. She has worked with numerous countries to identify and bridge the gap between their assessment, curriculum and pedagogy in relation to skills. In particular, she has worked with many developing countries to review and develop curriculum documentation, develop classroom-based assessment tools and generate teaching resources for skills including problem solving, collaboration, learning in digital networks, critical thinking, creative thinking and self-regulation. She has been an active lecturer for many years and has developed courses in both traditional formats and online formats, including both the design of the content and the assessment tasks. At the University of Melbourne, her Ph.D. identified a measurement methodology of twenty-first century skills through automated scoring and test delivery systems. Her experience in psychometric assessment, intervention work and research spans across the UK, Asia–Pacific and South America. She has undertaken consultancies for UNICEF, WorldBank, The Brookings Institution and UNESCO. Dr Scoular has been a Keynote Speaker on multiple occasions and recently served as a presenter and mentor to delegates at the G20 Youth Summit.

ORCID profile https://orcid.org/0000-0002-1858-3844

Introductory Analysis of the Rasch Model

Ian Storey

Abstract The author's encounter with the Rasch model came after a long experience teaching traditional statistical approaches, then being called in to advise a Ph.D. student on test items used for developing tests with a 'new' methodology: the Rasch technique. This technique develops efficient psychometric tests that produce interval measures. This is an astonishing feat, and it took the author some time to recognize that this was solidly based on probability theory, particularly regression. This experience, plus becoming involved in the Ph.D. supervision, led to an investigation of the theory behind the Rasch model, its 'philosophy', as well as the process of calibration, particularly the use of fitness measures. This kind of introduction to Rasch is certainly not rare. Furthermore, statistics is traditionally a hurdle for a large cohort of students, so perhaps the explanations given here benefit also from the teachers' perspective, even if potentially suffering because of it. The text therefore relates elements from probability theory. It also explains basic elements of psychometric testing, which is rarely covered in basic statistical texts. This text also provides a critique of significance and sample size as it relates to Rasch and, to some degree, to regression techniques in general. This subtle matter is perhaps appreciated by many, but it is briefly analysed here.

Rasch Model

The author has designed and taught higher education courses in statistics, and my first encounter with the Rasch model (RM) was perplexing. It should firstly be acknowledged that while probability theory has a firm foundation, 'textbook' statistics itself has had a rough past. At the foundation of modern statistics, Neyman, Pearson and Fisher engaged in bitter debates around the role and meaning of the p value, even questioning its impact on the 'intellectual freedom of the west'.

I. Storey (✉)
Lecturer Torrens University, Melbourne, Australia
e-mail: ian.storey@icloud.com

E. McKay (ed.), *Manage Your Own Learning Analytics*, Smart Innovation, Systems and Technologies 261, https://doi.org/10.1007/978-3-030-86316-6_3

> But while the rivals feuded … other researchers lost patience and began to write statistics manuals for working scientists. And because many of the authors were non-statisticians without a thorough understanding of either approach, they created a hybrid system that crammed Fisher's easy-to-calculate P value into Neyman and Pearson's reassuringly rigorous rule-based system. (Nuzzo, 2014, p. 151)

This friction carries over into the divisions between Bayesian and frequentist camps.

RM is part of this history, rooted in the theory of scientific measurement.

> [B]ecause proponents of the Rasch models emphasize the models' compatibility with fundamental measurement found in physical science, an analogy is made between how Kuhn explains the role of measurement in the physical sciences and how proponents of Rasch models explain the role of these models in the social sciences. In particular, these roles cannot be gleaned from textbooks in science and statistics, respectively. (Andrich, 2004, ppI-7)

With a very broad brush, we might say that RM is more about measurement than statistics. Nonetheless, it has firm roots in probability theory, particularly in logistic regression.

For someone unused to the Rasch technique, it helps, initially, to think of the testing instruments as tests composed of questions given to people who then answer the question 'correctly' or 'incorrectly'. But Rasch tests can be used more generally to measure latent traits. Questions need not have 'correct' answers, and indeed, they can have several responses.

The main advantage of a Rasch test is that it gives an interval measure of ability (or trait strength), and it does so very efficiently. Unlike classical test theory, which orders ability, Rasch gives a measure of ability. However, the test must be carefully developed beforehand. The test is calibrated in steps using metrics derived rigorously from probability theory. This process is the focus of someone wishing to develop tests using the Rash technique.

This chapter is a 'textbook' introduction to the mechanics of RM, explaining along the way the perspectives of the approach, hopefully without clouding the appreciation of RM.

Overview

From the very start, it should be explained that a 'test', in this context, is used for measuring a psychological trait. A test in RM is often distinguished from a hypothesis test by using the term 'test instrument'. (A test instrument is not meant as a replacement for a hypothesis test, and they have completely different applications.) It helps to visualize tests that contain 'correct' and 'incorrect' answers that measure test respondents' understanding or ability. People with higher ability will answer more questions correctly than those with lower ability. At the same time, questions with higher difficulty will tend to be answered correctly by less people than questions with lower difficulty.

Tests do not necessarily measure ability. They can measure latent traits in the 'human sciences, including educational achievement, human development, attitudes and health outcome' (Bond & Fox, 2001). They could attempt to measure any trait: driving awareness, attitudes to drug use, personality disorders, attitudes to social class, asthma severity, attitude towards the church or language ability. Almost any psychological trait. The questions, 'Do you prefer dogs to cats?' and 'Do you enjoy having your face licked by a dog?', do not necessarily have correct answers, but they could indicate a degree of difference in how much a person likes dogs.

Less prosaically, in a test to gauge a person's attitude to, say, drug use, some people may answer questions as if they are more tolerant of drug use while others are less tolerant. Again, a 'higher' response is not necessarily 'correct' in this case. The word "ability" is very commonly used in Rasch theory, although the more general term for the measured entity is trait or, more formally, latent trait (Bond & Fox, 2001, p 15).

Even so, it seems to aid visualization to imagine tests as measuring abilities: things such as mathematics ability, verbal skills, knowledge of thermodynamics. We will continue the practice here of referring to the measured entity as ability. In any case, ability is often quite literally what is being measured.

Tests are comprised of questions, more formally called 'items'. It is not unusual to hear either term being used. Answers to questions are often termed responses. Questions will often have just two possible response types. These are known as dichotomous questions or dichotomous items. We can think of the responses as being incorrect or correct. An incorrect response is generally coded as zero, while a correct response is coded as one. Some questions may have more than two levels of response, with higher 'scores' corresponding to higher levels of ability. A score of zero could be thought of as incorrect (or sometimes as not answered). A score of one is correct but does not display the same level of ability as a score of two and so on. Such questions are called polytomous. As Rasch discovered, mathematically, polytomous items are dealt with in almost the same way as dichotomous items.

The following correspondences between terms might help someone unfamiliar with RM to visualize tests, items and responses,

- Item ~ question,
- Individual, respondent, test subject ~ person,
- Trait, latent trait ~ ability,
- Test instrument ~ test composed of questions to be delivered to people,
- Response ~ answer to a question.

RM is grounded mathematically in the same kind of probability theory that fits data to a model, especially regression techniques. The population model is comprised of probabilities that a person with a given level of ability will be able to correctly answer a question with a given level of difficulty.

This seems familiar enough at first, but the model initially arrives at estimates of question difficulty and person ability, both at the same time. It seems to measure both people and questions in a side-by-side development. This seems like an impossibility, and in fact, it relies on some rather tidy mathematics, but it is nonetheless initially mystifying how this could be practically applied to the development of a measuring instrument.

The use of tests in RM is complex, but, for the purposes of this discussion, we can think of the process as being comprised of two basic steps: the development of a test instrument in a process known as calibration and then finally the use of the calibrated test to measure people's ability.

In the calibration stage, a test will change over time. The test will be delivered to different sets of individuals, several times. In each step, 'the' prospective test instrument will have some questions added and some removed. This process is designed to eventually produce a test instrument which has several properties. In the case of RM, a necessary property is unidimensionality. In very basic terms, this means that the test measures one thing. Tests are carefully developed over time to arrive at a test instrument that has this and other desirable properties.

This final test is another test comprised of a set of questions. However, there are many advantages of test calibration. For one thing, the test instrument does not need to be checked for fitness, and it is simply used as a measuring instrument.

The test gives reliable measures of relative ability across a wide range of skills. Compared to tests derived from classical test theory (CTT), a well-calibrated RM test can be quite efficient, needing only a relatively small number of questions for the same level of test discrimination. The test also produces an interval measure (as explained below). Such a test can be used as a measuring instrument. In fact, the test can be reliably used with just one person sitting the test.

The analogy to a ruler is widely used. Photographs in texts, Web pages and videos show physical rulers to emphasize the point that a ruler, once developed through calibration, can be given to anyone to use. Conference presenters have been known to distribute rulers to emphasize the point. But tests need to have been well calibrated before they can become a reliable 'ruler' for the trait we wish to measure.

For the remainder of this chapter, we focus on the 'tests' as used within the calibration process. Each of these tests in the series can be thought of as a prototype for the final test—the test instrument.

Tests

The calibration process is focused on deriving a good set of questions and on determining their difficulty. In this process, it is unavoidable that we also need to determine respondents' abilities. Even so, it is important to remember that determining the question difficulty is the focus of calibration. At this point, the respondent's ability measures might not be accurate. However, the pragmatic goal is to develop a test instrument (a ruler) which accurately measures respondent's ability.

In a single test, suppose there are N individuals (people) and L items (questions). Suppose the questions are dichotomous, and for the purposes of this discussion, a 'correct' response is coded as 1, and 'incorrect' is coded as 0. Polytomous questions (having more than 2 response states) can also be modelled using higher numbers.

Consider the nth person answering the ith question on the test. According to the Rasch model, the person has a certain level of ability, and let us represent it as θ_n. The ith question has a certain level of difficulty, call it δ_i,

- θ_n: ability of the nth person,
- δ_i: difficulty of the ith question.

In RM, the units for these measurements are *logits*, as will be explained.

A test is treated as a random experiment from standard probability theory. The data consists of responses. A person's answer to a question corresponds to a response. Each response represents a random event. In practice, there are often missing responses. Let us suppose, however, that ω_{ni} is the unique outcome corresponding to individual n's response to item i (the answer given by person n to question i), and it is either correct or incorrect. We designate the random variable X_{ni} to be the indicator of whether the answer is correct,

- $X_{ni} = 0$ if the question was answered incorrectly and
- $X_{ni} = 1$ if the question was answered correctly.

This is a Bernoulli random variable, and the probability density function (pdf), $f_{X_{ni}}$, is the probability of answering the question either correctly or incorrectly,

$$f_{X_{ni}}(1) = P(X_{ni} = 1)$$
$$f_{X_{ni}}(0) = P(X_{ni} = 0) = 1 - P(X_{ni} = 1)$$

If person n has a high ability, and if question i is easy, $\delta_i \ll \theta_n$, then the probability of a correct answer will be high (close to one).

$$P(X_{ni} = 1) \approx 1$$

On the other hand, if the individual's ability is low and the question's difficulty is high, $\theta_n \ll \delta_i$, the probability of a correct response is low (close to zero).

$$P(X_{ni} = 1) \approx 0$$

In RM, if the ability of the person is exactly matched with the difficulty of the question, $\theta_n = \delta_i$, we would expect an equal probability of correct and incorrect response,

$$P(X_{ni} = 1) = 0.5$$

In this model, the probability of success is dependent only on the difference between person ability and question difficulty, $\theta_n - \delta_i$. This model, if it applies, has enormous advantages. If it calibrates well, it gives confidence that the test (ruler) can be used for measuring a single, independent trait. That is, the measurement has unidimensionality.

In the Rasch model, both item difficulties and individual abilities can be measured along a meaningful interval scale. The relative ability of individuals depends on the size of the difference in logit scores. This is an important element of RM.

In classical test theory (CTT), the result is measured using a raw score or, equivalently, the proportion of correct responses. If individual n has greater ability than individual m using the CTT score, then we would expect individual n to have a higher score than individual m using the CTT score (accounting for random variations). The CTT score is an ordinal measure. That is to say, it rates ability by order, from lower to higher.

The Rasch measure is an ordinal measure, but it is also an interval measure. The size of the differences between the scores is indicative of difference in ability. For example, a person with a score of 3 logits has twice as much more ability than a person of score 1 logit as a person with a score of 2 logits (within limits of accuracy). When carefully calibrated, the measuring instrument developed by IRT supplies meaningful interval data, not just ordinal data.

However, the scale is not a ratio scale. That is, there is not a natural zero in the scale. A person with a score of 2 is not twice as good as a person with a score of 1. By convention, a test is usually set with an average item difficulty of zero. But this is merely a convention.

So, with the Rasch model the zero point is immaterial. What is not arbitrary is the intervals. In explaining Rasch measurement, this is often compared to temperature measurement where the zeroes of Celsius and Fahrenheit are different (this might not be the best possible example since there is an absolute zero temperature, and the scales have different interval lengths, but the explanation seems to be helpful).

Test scores will usually have values that lie roughly between $--3$ and -3. We have the following rough characterization of item difficulty,

- Very easy: -2.6
- Easy: -1.5
- Medium: 0
- Hard: 1.5
- Very Hard: 2.6.

Ability measures, θ_n, and question difficulties, δ_i, are initially derived in the one process. We develop a sense of a person's ability by looking at the difficulty of the items they were able to answer correctly. At the same time, we develop a sense of the difficulty of a question by looking at the abilities of people who were able to answer the question correctly.

In fact, the processes of finding individual ability and item difficulty are very similar; they can, in some approaches, be mathematical duals of each other. In any case, the question becomes 'How are the values for δ_i and θ_n initially obtained?'.

We begin with, models for the probabilities (probability density function, pdf) are associated with the random variable, X_{ni}. The formula in Eq. (1) is the one-parameter logistic (PL1) model using just one parameter for question difficulty, δ. This model applies the logistic function,

$$f_{X_{ni}} = P(X_{ni} = 1) \triangleq \text{logistic}(\theta_n - \delta_i) = \frac{e^{\theta_n - \delta_i}}{1 + e^{\theta_n - \delta_i}} \tag{1}$$

This is the main model discussed in this text. The probability for an unsuccessful response is simply one minus the above value,

$$P(X_{ni} = 0) = 1 - P(X_{ni} = 1) = \frac{1 + e^{\theta_n - \delta_i} - e^{\theta_n - \delta_i}}{1 + e^{\theta_n - \delta_i}} = \frac{1}{1 + e^{\theta_n - \delta_i}}$$

In the following equation, it might be thought that the expression on the right is the more 'simplified' version of the expression used to define the logistic function.

$$\frac{e^{\theta_n - \delta_i}}{1 + e^{\theta_n - \delta_i}} = \frac{1}{1 + e^{-(\theta_n - \delta_i)}}$$

The expression on the left, however, can be more convenient in some contexts and is often used as the definition.

The logistic function is the inverse of the logit (logodds) function defined as

$$\text{logit}(p) \triangleq \ln \frac{p}{1 - p}$$

And we have

$$\text{logit}(P(X_{ni} = 1)) = \theta_n - \delta_i$$

As with regression (linear and logistic), approximate parameters are found by fitting test data to the model, minimizing a cost measure of the difference between the data and the model. This process of optimization is discussed later.

Test outcomes are usually notated using a lower-case x, as is standard with sample data (x_{ni} will be either 0 or 1 in the dichotomous case). In the literature, estimated values of population parameters are sometimes written using a 'hat' over the value, $\widehat{\theta}_n$ and $\widehat{\delta}_i$, as is standard in regression analysis. Sometimes, letters from the Roman alphabet are used for estimates, sometimes B_n for ability and D_i for difficulty.

In the logistic formula of Eq. (1), it is clear that when ability equals difficulty, $\theta_n = \delta_i$, the logistic function returns a value of 50%,

$$P(X_{ni} = 1) = \frac{e^{\theta_n - \delta_i}}{1 + e^{\theta_n - \delta_i}} = \frac{e^0}{1 + e^0} = \frac{1}{2}$$

When the individual's skill, θ_n, exactly matches the item difficulty, δ_i, there is a 50% chance that the individual will give the correct response. When the skill level greatly surpasses item difficulty, $\theta_n \gg \delta_i$, say $\theta_n - \delta_i = 5$, the chance of success is nearly one,

$$P(X_{ni} = 1) = \frac{e^{\theta_n - \delta_i}}{1 + e^{\theta_n - \delta_i}} = \frac{e^5}{1 + e^5} = 0.999955$$

Someone with an ability of 2.5 logits will almost always correctly answer a question of −2.5 logits. Similarly, when the item difficulty surpasses the individual's skill level, say $\theta_n - \delta_i = -5$, the chance of success becomes negligible,

$$P(X_{ni} = 1) = \frac{e^{\theta_n - \delta_i}}{1 + e^{\theta_n - \delta_i}} = \frac{e^{-5}}{1 + e^{-5}} = 0.000045$$

Someone with an ability of −2.5 logits will almost always fail to correctly answer a question of 2.5 logits.

In this model, it is assumed that the item difficulty value, δ_i, is independent of an individual's skill, θ_n. The probability of success is dependent only on the difference between the two. This conforms to the requirements of unidimensionality. The probability is not dependent on any other factor,

Unidimensionality: $P(X_{ni} = 1|\theta_n, \delta_i, \varphi) = P(X_{ni} = 1|\theta_n, \delta_i)$. The response probability does not depend on variables other than person ability and item difficulty (Hatzinger, 2010).

If a test is not unidimensional, it is not a reliable measure of a single given trait. In the process of calibration, test items are developed over many steps to ensure, as much as is possible, that unidimensionality holds. Many of the initial steps are exploratory in nature, so the danger of post hoc analysis is small. Also, tests in the process of calibration are designed to be independent of each other, at least in terms of respondents, as much as is possible.

Once a test has been fully developed, the (relative) item difficulties, δ_i, are known. Then, these test items can be used to measure individual's ability. They can be used to determine the values of θ_n in a test, but they can even be used to test the ability of just one individual alone, θ. As discussed earlier, they can be used like a ruler.

The probability density function of individual ability, θ, as a one-parameter function, is known as the item response function (IRF),

$$f_i(\theta) = \frac{e^{\theta - \delta_i}}{1 + e^{\theta - \delta_i}} \quad (2)$$

Note that this is distinct from Eq. (1) since it is a continuous function of the single parameter, θ, while δ_i is a 'constant'. This is also known as the item characteristic function (ICF). The curve for this function is known as the item characteristic curve (ICC). It has a distinctive sigmoid ('S') shape (see Fig. 1).

Given the apparent symmetry between items and individuals, we could define a complementary function to the function of difficulty. This is known as the person response function (PRF) (Walker et al., 2018, p 50). The IRF has been given much more attention than the PRF because it is heavily used in assessing items in the calibration process. Even so, it has been argued that PRF is a useful measure of misfit (Walker et al., 2018).

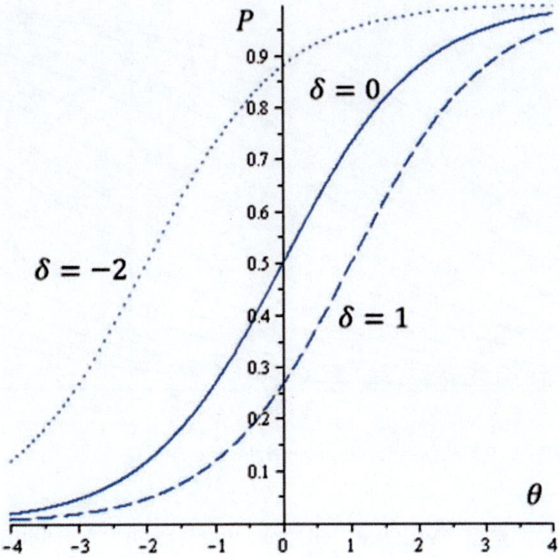

Fig. 1 ICC curves; dashed line with $\delta = 1$ represents most difficult item

The IRF function gives a snapshot of the likelihood of a correct response as a function of the individual's ability. Figure 1 shows three ICC curves. Each curve represents a different question with a different level of difficulty: −2 logits, 0 logits and 1 logit. The item with $\delta = 1$ logit (dashed line) has the greatest difficulty. It has consistently lower probabilities of being answered correctly than the other two. People find it harder to answer this question correctly than the other two. A person with ability $\theta = 0$ has a 50% chance of correctly answering the question with $\delta = 0$ logits (solid line). On the other hand, the dotted graph represents an easy question, with difficulty $\delta = -2$ logits. People with abilities of 0 or greater have a better than 88% chance of answering such questions correctly.

ICC curves are shifted horizontally according to the difficulty of the question. Harder questions are shifted to the right; less people will answer them correctly, and the probabilities are lower for the same abilities. And easier questions are shifted to the left because probabilities are higher for the same abilities.

For a good measuring instrument, in terms of RM, all the items (questions) in the test should have similar ICC shapes. However, a test is not very well calibrated if all questions measure the same level of ability. To cope with a range of individual abilities, items should have a wide range of difficulties and an even spread of difficulties.

The property of a test in which questions have similar sigmoid shapes shifted left or right is known as uniform differential item functioning (uniform DIF). An ideal test would consist of many questions, spanning the entire range of difficulty (a span of around 6 logits), as depicted in Fig. 2. Tests in practice will have more than 20 items. Such a test should cover the entire range of difficulty, matching abilities of individuals likely to take the test in future.

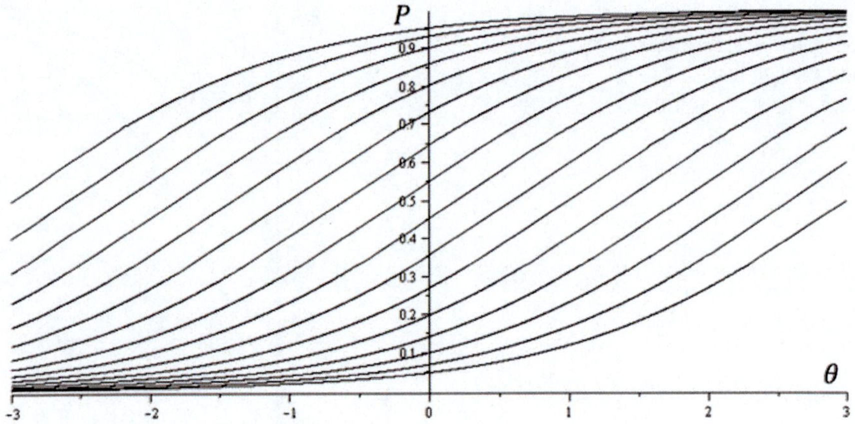

Fig. 2 ICC curves for ideal general test

Not all useful tests will have uniform DIF. A test might sometimes have questions that target just one level of ability. Such a test might be used for testing whether candidates are worthy of taking on a new role, of gaining certification for a business standard or of passing a test in understanding a piece of theory. Such a test would result in either pass or fail. But the real challenge with Rasch measurement is to derive a test which results in an interval scale of measure of respondents' abilities.

A recent development is to deliver test questions by computer and to quickly zero in on the respondent's level of ability. (There will be a kind of binary search for a respondent's ability, although of course with some redundancy to account for random effects.) The challenge is to keep the number of questions to a minimum while still deriving a reliable result. This is known as computer adaptive testing (CAT), also called computerized adaptive testing (Magis et al., 2017). Even with such a test, a bank of questions with a wide range of difficulties is needed.

Tests are more effective if the range of question difficulties matches the range of individual ability. It is exceedingly unlikely, however, that there will be an exact match of average question difficulty and average individual ability. Consider, for example, the histograms in Fig. 3 representing test outcomes. By convention, the average item difficulty is set at zero logits (the test instrument is 'zeroed' against questions rather than against potentially different sets of individuals).

Figure 3 shows software output of test analysis with two histograms drawn back-to-back, one for the individuals (at the top) and one for the items (at the bottom). (Histograms count frequencies of outcomes, with outcomes placed inside various ranges of values, called 'cells' or 'bins'.) In this test, the average item difficulty is set at zero logits. However, in this case, the average person produces a score of 0.701 (reported as mean 0.701 at the top right of the output). The average item difficulty, in this test, is lower than the average person ability.

If there is overlap between the ranges of items or individuals, then either questions are being wasted, because they are either too easy or unanswerable, or people are not

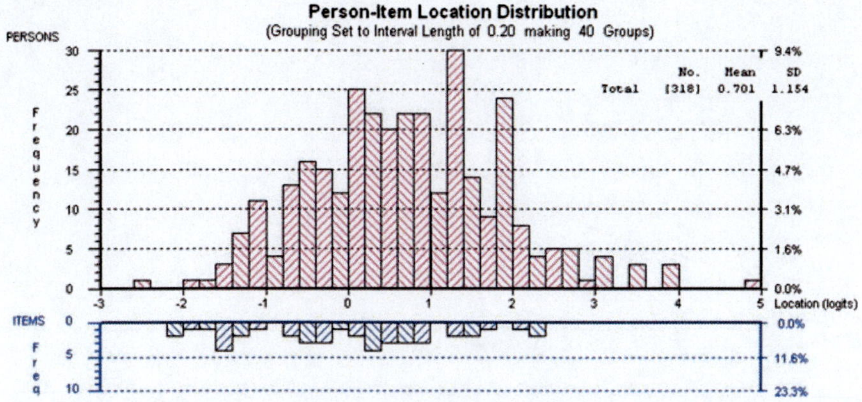

Fig. 3 Person-item location distribution (Wikipedia, 2021)

being properly tested, because there are no questions to match their ability levels. For example, in Fig. 3, there are quite a few people who have a higher ability than the highest question difficulty. Adding harder questions to the test will help to better distinguishing those individuals with higher ability.

Rasch assumes the model of Eq. (2) (for dichotomous items). Other models are used in IRT. The following is a more general three-parameter logistic model (PL3).

$$p_i(\theta) = c_i + \frac{1 - c_i}{1 + e^{-a_i(\theta - b_i)}} \tag{3}$$

This model has three parameters for each item: a_i, c_i and b_i. Compare this with the PL1 model (Eq. 2), which uses only the one 'parameter', δ_i.

In the PL3 model, there is a minimum probability for a correct answer, c_i. Refer to Fig. 4. This kind of function can accommodate guessing in a multiple-choice question, where we might expect a minimum correct response rate of say $c_i = 25\%$, if the question has four options. If the question is open-ended, with a small chance of an accidental correct response, then $c_i = 0$, and we have the common form of the two-parameter logistic model (PL2).

Consider the ICC curves for the three items shown in Fig. 5. These follow the PL2 model ($c_i = 0$ in Eq. 3). Each of these items has the same level of difficulty, with an individual with ability $\theta = 0$ (in this scale) having a 50% probability of correctly answer any of the three items. However, for an ability level slightly above zero, the item represented by the solid line will have a higher chance of being answered correctly than the other two items. It will better 'discriminate' between abilities above and below the average. The item represented by the dotted line will be virtually unable to discriminate abilities above and below zero.

The slope of the line at the 50% point is known as the discriminant. The item represented by the solid line has the largest discriminant, $a = 4$, and the item represented by the dotted line has the smallest discriminant, $a = 0.25$.

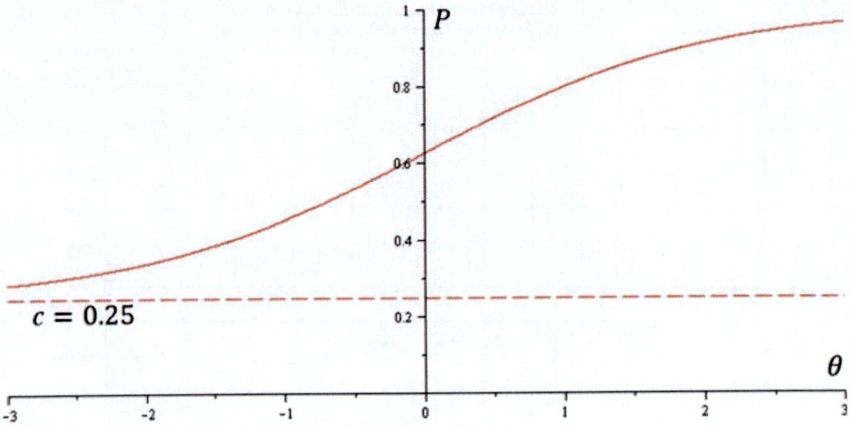

Fig. 4 3PL model

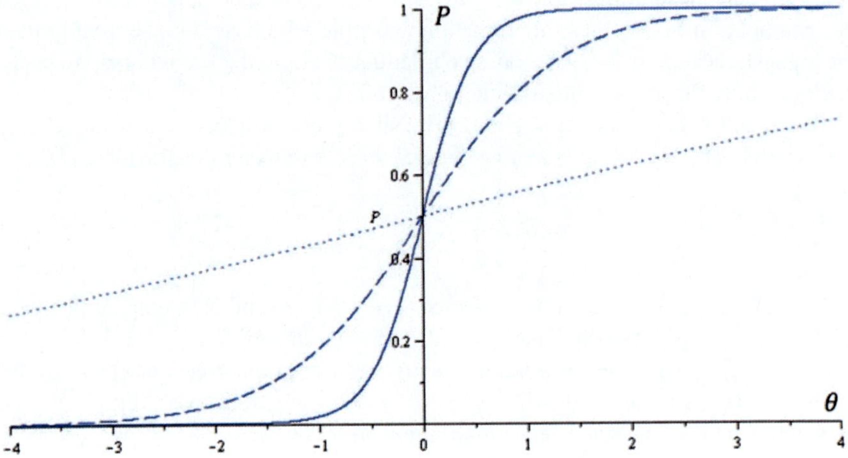

Fig. 5 Items with varying discriminants, $a = 4$ (solid), $a = 1.5$ (dashed line), $a = 0.25$ (dotted line)

An item with an extremely low discriminant is clearly not particularly useful. It is inefficient at 'discriminating' between individuals of different abilities. A test composed of such questions would require many more questions than one with questions with higher discriminant value to obtain the same level of reliability.

It might be thought then that items with the largest discriminants would be the most valuable in a test, since they give a more certain response for individuals with abilities either side of the item difficulty. However, there are several reasons why such a question is not desirable. Firstly, for the purposes of producing a test that is unidimensional, an item with a high discriminant is likely to be measuring some other trait than the one intended. Also, if the discriminant is high, the item can distort the overall test information profile. The inclusion of items with a high discriminant will tend to 'push out' items with a close 'δ' value. A high discriminant is not as bad as a low discriminant, but if it is too high, it should be regarded with suspicion.

Typically, in designing a test instrument, items with discriminants that are 'too high' or 'too low' will be removed from consideration. Note that this is a post hoc process and so should be performed with exploratory sets of data. It is best too, if fresh respondents can be applied in separate tests as much as is feasible, although this is not strictly necessary with exploratory data sets.

The tests described in this section have had dichotomous items. That is, they have used items with two possible responses. The extra levels are treated in very much the same way as dichotomous items.

Parameter Estimation

As with parameter estimation generally, estimation of the test parameters, item difficulty and person ability, is performed by optimization of a fitness measure or fit statistic. A fit statistic is some measure of the 'distance' of the average values predicted by a model (expected values) from sample data (observed values). In theory, the fitness measure should measure the distance to the population model, but this is almost never known, so fitness is usually a measure of the 'distance' of collected data to a model.

The 'distance' between the model and the data becomes the cost function for the optimization process. For example, in linear regression the cost function is the sum of squares of differences between observed data points, O_i, and expected data values, E_i,

$$c = \sum_i (O_i - E_i)^2$$

In the process of estimating parameters, Rasch is essentially a form of logistic regression. As with logistic regression generally, the optimization process does not admit a neat closed solution (unlike linear regression). Numerical methods are usually needed to find the optimum.

With small 'toy' problems, involving say five items and two individuals, it is possible to use the Newton–Raphson method to closely approximate the maximum likelihood estimate of parameters. The process is fast and essentially accurate to the calculator's degree of accuracy.

The responses in a test are assumed independent, so the likelihood of any given set of data is the product of the probabilities of each response,

$$\begin{aligned} &P(X_i = O_i \text{And} \ldots \text{And} X_m = O_m) \\ &= P(X_i = O_i) \times \ldots \times P(X_m = O_m) \end{aligned}$$

(Usually, the log likelihood is used, which makes the problem slightly simpler.) We find the parameters which maximize this probability. (The 'distance' between the model and data is minimum with maximum likelihood.) This method can be used to estimate respondent abilities if item difficulties are known beforehand, which is known as the maximum likelihood estimate (MLE). Or it can be used to estimate both item difficulties and individual abilities at the same time, which is clearly a more complex problem. This is known as the joint maximum likelihood estimate (JMLE).

Unfortunately, with large numbers of items and/or respondents the maximum likelihood problem is nonconvex. That is, there may be many local maxima, as shown in Fig. 6b. This can make the optimization process quite unstable; the Newton–Raphson will find local minima, but not necessarily a global optimum.

Optimization techniques can make special use of the properties of Rasch probability functions. It is also possible to use cost functions that make the optimization, or at least initial optimization steps, faster and more likely to reach the global optimum. The use of special cost functions to improve optimization is a technique often employed in logistic regression outside the Rasch model. In fact, different Rasch software packages can use different optimization techniques, sometimes producing slightly different results.

Fortunately, the complex process of optimization is today usually left up to software, although optimization was performed by 'hand' in the past (using tables and/or desk calculators). With larger sets, of course, dedicated software is more reliable. The practitioner is generally not interested in the exact details of the software.

However, the matter of fit statistics does not end with model estimation; in fact, they are a fundamental part of the wider calibration process. It is useful to have some understanding of the various fit statistics used in calibration. In this text, we examine the two main fit statistics, infit and outfit.

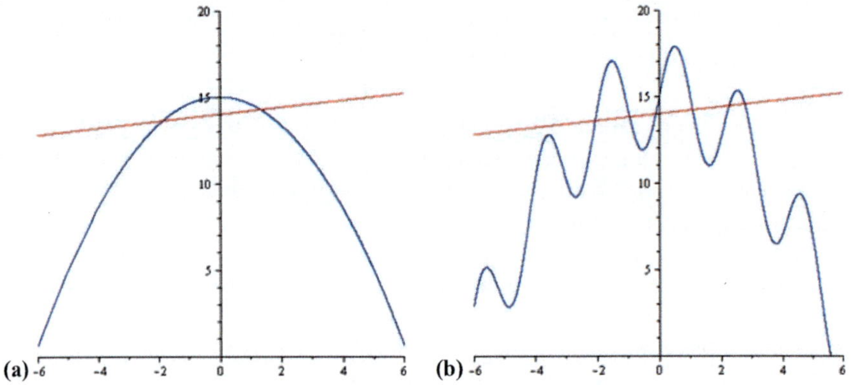

Fig. 6 **a** Convex problem and **b** nonconvex problem

Fit Statistics

Measures of the extent to which sample data conforms to a model are widely used in statistics. They are used to express confidence in the model or to show the significance of the model using standard hypothesis testing techniques. For example, the χ^2 goodness-of-fit statistic is widely used to test if a sample's data follows a given distribution. Cost functions used in the process of parameter estimation are also forms of fit statistics. In RM, fit statistics are heavily used in item calibration. '*[I]tems are discarded which do not satisfy the criteria considered important from the point of view of the model. For this sample-free model the essential criterion is the compatibility of the items with the model*' (Wright & Panchapakesan, 1969, p 25).

There are many different fit statistics used in RM. Different fit statistics are sensitive to different violations from the model (Wu & Adams, 2007). Here, we discuss the two most common measures, infit and outfit, and we briefly discuss differential item functioning (DIF).

χ^2 *Statistic*

Take a single data point (a single individual's response to a single question), with a random variable X_{ni} representing the score from the test response of individual n to test item i. In the dichotomous case, X_{ni} is either 0 or 1. The following response statistic, known as the standardized residual, has an expected value of zero and a standard deviation of one,

$$z_{ni} = \frac{X_{ni} - E_{ni}}{\sigma_{ni}} \tag{4}$$

Here, E_{ni} represents the expected value of X_{ni} and σ_{ni} is the standard deviation, using the proposed probabilistic model. In the dichotomous case, the response is modelled as a Bernoulli trial, and the expected value equals the probability of success, calculated using Eq. (1). The variance is calculated in the usual way for a Bernoulli trial,

$$\sigma_{ni}^2 = p_{ni}(1 - p_{ni}) = E_{ni}(1 - E_{ni})$$

Polytomous data may have some advantages over dichotomous as far as fit measures are concerned (Smith et al., 2008), but in this section we confine ourselves to dichotomous data.

The sum of squares of a standard normal, with residuals that are normally distributed, follows a chi-square (χ^2) distribution. If the test is large enough, we can appeal to the CLT and we might suppose that the statistic z approximates a standard normal distribution. This may be problematic given the wide divergence of the Bernoulli distribution from the normal, but let us suppose for the moment CLT allows the approximation.

There is one value of z_{ni} for each individual-item response in any given test. The χ^2 value is then derived from the sum of z_{ni}^2. Sums can be taken by individual. That is, summing over items responded to by individual n we have the χ^2 goodness-of-fit statistic,

$$\chi_n^2 = \sum_{i=1}^{L} \frac{(X_{ni} - E_{ni})^2}{\sigma_{ni}^2}$$

In an already calibrated test, this statistic gives useful information about an individual's response. On the other hand, we can sum over all individuals' responses to a given test item,

$$\chi_i^2 = \sum_{n=1}^{N} \frac{(X_{ni} - E_{ni})^2}{\sigma_{ni}^2}$$

In the process of test calibration, this gives useful information about a test item. Here, we focus on the sum over items for individual n, but much of what is said also applies to the sum for an item.

Note that the χ^2 statistic follows the χ^2 distribution if the sample events, the responses, are independent of each other. The separate questions should not logically follow from each other. Getting one question right should not make it easier to correctly answer another. One question should not depend on the detail of another. Unfortunately, the questions should not teach the student as they go, as you might find in a well-designed set of pedagogical questions at the end of a chapter.

Outfit and Infit

In the output of Rasch analytical software, the χ^2 value is generally converted to a mean square value. That is, it is multiplied by a constant to give an overall expected value of one. (Issues of degrees of freedom being sacrificed by the determination of model parameters are ignored here).

Assuming that the separate responses are independent, the expected value of χ^2 is

$$\mathbb{E}[\chi_n^2] = \sum_{i=1}^{L} 1 = L$$

So, the mean square measure is

$$\text{Outfit}_n = \frac{1}{L}\sum_{i=1}^{L}\frac{(X_{ni} - E_{ni})^2}{\sigma_{ni}^2}$$

This statistic is termed outfit or sometimes unweighted mean square. We have outfit scores for both individuals and items. The outfit score for item i is given as

$$\text{Outfit}_i = \frac{1}{N}\sum_{n=1}^{N}\frac{(X_{ni} - E_{ni})^2}{\sigma_{ni}^2}$$

Under the assumption that the z_{ni} approximately follow a standard normal distribution, the χ^2 statistics should follow a χ^2 distribution. However, the outfit measure is extremely sensitive to outliers—someone with high ability making a mistake on a very easy question or a difficult question guessed correctly. In the case of a dichotomous question (a Bernoulli trial), there are only two outcomes and the z score for a given response is

$$z = \frac{X - E}{\sigma} = \frac{X - p}{p(1 - p)} \tag{5}$$

When the outcome is successful, and the individual's ability is much higher than the question difficulty, we would expect that the individual would be successful. For example, an individual with an ability of 2.5 logits answering a question with a difficulty of −2.5 logits would be expected to answer the question correctly with a probability of

$$p = \frac{e^5}{1 + e^5} = 0.9933$$

This is almost one. If the respondent answers the question correctly, $X = 1$, the observed response is almost equal to the expected value, p. Consequently, the z^2 value is very low,

$$z^2 = \frac{(1 - p)^2}{p(1 - p)} = 0.0067$$

However, if the respondent makes a mistake, as happens from time to time with human beings, and the question is answered incorrectly, $X = 0$, we have an extremely large value for z^2,

$$z^2 = \frac{(0 - p)^2}{p(1 - p)} = 148.4$$

This value heavily skews the overall χ^2 sum. According to the model, the probability of such an error is extremely small, $(1 - p) = 0.006693$, or around one in

150. The outfit score has an expected value of 1, however, if this response discussed earlier, with a value of $z_{ni}^2 = 148.4$, is included in a total of, say $L = 30$, items, then the outfit score for the individual would be increased by

$$\frac{z_{ni}^2}{L} = 4.95$$

One response would greatly inflate the outfit score.

But we could imagine that, under the pressure of a test, a respondent could have a momentary lapse of concentration, or they could simply misinterpret the question. People with skill will sometimes make mistakes that have little to do with their level of ability on the item being tested. This can unfairly inflate the overall χ^2 value and hence the outfit score. On the other hand, people will also sometimes make correct educated guesses. The Rasch model assumes '*that the effect of guessing is negligible. Our experience with the analysis of real data suggests that the model is quite robust with respect to departures from these assumptions*' (Wright & Panchapakesan, 1969, p 25). Experience shows that the model produces useful test instruments.

Note in passing that if the item discrimination is 'too good', if it is much smaller than one, then almost all responses with probability greater than 50% are answered correctly and almost all responses with probability less than 50% are answered incorrectly. In this case, the infit score for the item will be exceedingly small. Initially, it might be thought that this is a good thing, and certainly, it is not as bad as having a score which is too large. However, it may indicate that the item is indicating something other than what is intended. Certainly, it indicates departure from the population model.

Figure 7 shows a plot of z^2 for 'unexpected responses' (success for questions with negative values for $\theta - \delta$, and fails for questions with positive $\theta - \delta$). Clearly, the outfit measure is heavily skewed for outliers, at the 'outer edges' of the logit differences.

To ameliorate the sensitivity of outfit to outliers, a weighted score can be adopted. The individual scores can be weighted by "information". In the dichotomous Rasch model, the information weighting function is given as

$$\sigma^2 = pq = p - p^2$$

This is largest for points where the expected outcome is close to 0.5, but it approaches zero at either extreme: where the expected probability is either zero or one. Thus, we have a discounting effect for the extremes.

Weighting the standardized data value, z, by information, σ, gives a squared term of the form

$$(z\sigma)^2 = \frac{(X - E)^2}{\sigma^2}\sigma^2 = (X - E)^2$$

This is plotted for 'unexpected responses' in Fig. 8. The score increases for larger

Fig. 7 z^2 for 'unexpected responses'

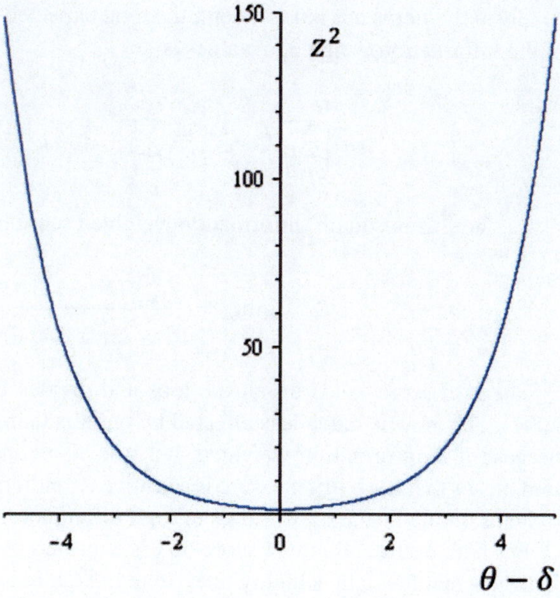

Fig. 8 Information weighted z^2 for unexpected responses

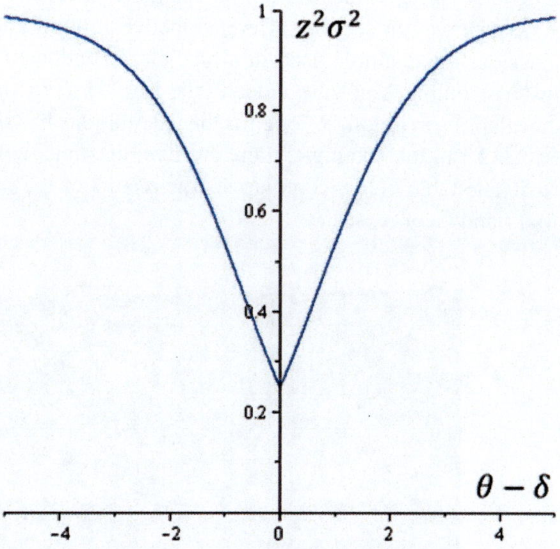

excursions from expected responses, but it does not increase without limit. Instead, for outliers it approaches a maximum of one. The weightings for outliers are no longer in the range of tens or hundreds but are now less than one. If the information weighted values are summed, then the impact of outliers is greatly reduced.

Given the items are independent, the total expected value for the sum of squares of the information weighted z values is

$$\mathbb{E}\left[\sum_{i=1}^{L} z_{ni}^2 \sigma_{ni}^2\right] = \sum_{i=1}^{L} \sigma_{ni}^2 \mathbb{E}\left[z_{ni}^2\right] = \sum_{i=1}^{L} \sigma_{ni}^2$$

So, the ‘mean square’ information weighted score is given as

$$\text{Infit}_n = \frac{\sum_{i=1}^{L} (X_{ni} - E_{ni})^2}{\sum_{i=1}^{L} \sigma_{ni}^2}$$

The infit score ‘adds up all the tops and divides by all the bottoms’ (Linacre, 2001). This now is much less affected by outliers than outfit. This is known as infit, because it is information weighted, but it is also convenient that the term ‘outfit’ reminds us that the outfit measure is sensitive to outliers, while infit is more sensitive to items on the inside of the range of logit differences.

For both outfit and infit, a score of 1 is expected. A score much greater than one indicates that the data is highly unexpected. A score less than one indicates that the data fits ‘too well’ (the variances around expected values are smaller than the model). Of course, these values are random statistics that have their own variations.

A nice way to see the difference between outfit and infit is with a toy example. Consider the example used in a nice class conducted by Linacre (2001), recorded in three online YouTube videos (see Fig. 9). (For those who wish to follow the calculations in detail, R code for the calculations has been supplied in the appended Sect. 12.) In this example, there are five questions with difficulty levels, δ_i, from 1 to 5. The table below summarizes the results of the test run with two respondents, Mary and George.

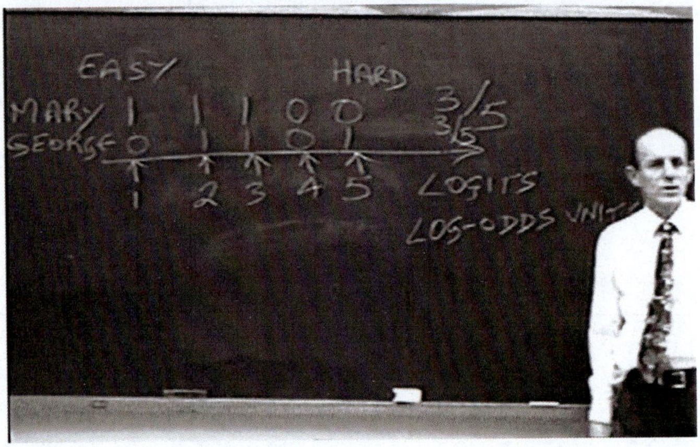

Fig. 9 John Michael Linacre’s example

Mary	1	1	1	0	0
George	0	1	1	0	1
Question difficulty (logits)	1	2	3	4	5

Mary succeeds on all questions with a difficulty between 1 and 3 logits. She fails on all questions with a greater difficulty. Mary's responses follow the expected Guttman pattern (Bond & Fox, 2001). If Mary has an ability level between 3 and 4 logits, all responses are the most likely responses to the questions. Mary and George have the same raw score, 3 out of 5, but George's responses do not follow the Guttman pattern. If we suppose his ability level lies between 3 and 4 logits, then there are two 'unlikely' responses or outliers. The easiest question has a difficulty level of 1, yet George makes a mistake with this question. This clearly qualifies as an outlier. He also answers the hardest question correctly.

The probabilities of these responses depend on the ability level of the respondents. It turns out that the maximum likelihood estimate of ability is the same for both respondents and works out to be $\hat{\theta} = 3.591$ logits (readily found using Newton–Raphson). With this ability, the probabilities for the responses found, using the Rasch model, are as shown in the table below. This table also includes the variances for each response. The probabilities for the unexpected responses are shown in bold.

Item	1	2	3	4	5	Λ
Pr, Mary	0.930	0.831	0.644	0.601	0.804	0.240
Pr, George	**0.070**	0.831	0.644	0.601	**0.196**	0.004
σ^2	0.065	0.141	0.229	0.240	0.158	

The overall maximum likelihood (the product of the individual probabilities) for Mary's response, 0.2402, is much higher than George's, 0.0044. George's two unexpected responses (his responses to the first and fifth items) have very low probability. Hence, his response has a very small overall likelihood. Nonetheless, the ability estimate, $\hat{\theta} = 3.591$ logits, gives the maximum likelihood for both Mary's and George's responses.

Because of unexpected responses in George's case, the sum of squares of observed minus expected responses is much greater for George. The χ^2 value for George is 18.9, while χ^2 for Mary it is 1.74. The table below shows the contributions of each response to the χ^2 sum, using z^2 where z is calculated using Eq. (5).

Item	1	2	3	4	5	χ^2
z^2, Mary	0.075	0.204	0.554	0.664	0.244	1.74
z^2, George	**13.340**	0.204	0.554	0.664	**4.093**	18.9

George's unexpected responses, especially the mistake with the very easy first item, make an overly large contribution to the overall χ^2 statistic. This gives outfit values of 3.77 and 0.348, respectively. There is an almost 12 to 1 ratio between the two values.

The outfit values indicate that George's response is not a good fit to the Rasch model, but it is perhaps an overly large impact for just two unexpected responses. The infit values for the responses are 2.19 for George and 0.431 for Mary. There is a smaller ratio for these values of around five to one, indicating that the outliers have been given a smaller weighting.

The outfit measure is proportional to the χ^2 statistic (multiplied by the number of degrees of freedom). In the past, calculations of χ^2 cdf probabilities have relied on the quite accurate approximation suggested by Wilson and Hilferty (1931). This transforms the cdf of the χ^2 distribution into that of a standard normal distribution, allowing practitioners to look up values in tables of the standard normal in the 'back of the book'. Most computer statistical packages will calculate χ^2 distribution values accurately. Nonetheless, the Wilson–Hilferty approximation is available and it is quite accurate. It is also still heavily referenced in the literature, usually called a t statistics, although perhaps it is better termed a standard normal, z, statistics (Linacre, 2020). In any case, its function is to approximate the cdf for the χ^2 distribution.

As a rule of thumb, *"Since the fit t statistic can be regarded as a normal deviate, a t value outside the range of* -2.0 *to* 2.0 *(or* -1.96 *to* 1.96*, to be more precise) can be regarded as an indication of misfit, at the* 95% *confidence level"* (Wu & Adams, 2007, p82). This is simply a mapping between the χ^2 and normal distributions, with the consequence that the model is incorrect with 5% significance.

Significance and Sample Size

As with the χ^2 statistic, badly fitting population models tend to produce large outfit and infit statistics. However, interpreting this in terms of p values for the purposes of item calibration can be misleading.

We have already discussed the reasons that the outfit measure may be slightly inflated. Another difficulty is that p values are used to strongly affirm the alternative hypothesis, not the null hypothesis. In our case, we would like to affirm that the Rasch model applies to the population. However, this is the null hypothesis in the case of the χ^2 test. This same issue arises in all regression techniques where χ^2 is used to test the assumptions of the model, including linear and logistic regression, SEM, etc.

The way that the χ^2 test applies in this context is the opposite, in a sense, of what is usual in a hypothesis test. We do not affirm with 'significance' that the model is correct. We simply affirm that it is not incorrect. The double-negative logic can seem convoluted. In the analogy with a court case, a large p value does not prove the defendant innocent; but it does not prove that he is guilty.

So, a low outfit or infit score does not show that the data fits the model. Rather, it shows that we do not have good reason to disbelieve the model. There is a relatively good chance that a somewhat badly fitting item will 'pass' the goodness-of-fit test in some cases. By this reasoning, if the test item is used in a later test, with a new set of respondents, and has an overly large fitness measure, then the item should be regarded as suspect, even if it 'passed' previously.

Suppose as an example that we have an outfit statistic of 1.4 for a test item with $N = 30$ respondents. What is the p value for the null hypothesis that the data does fit the Rasch model? That is, what is the probability that the outfit statistic would be greater than 1.3 given that the model is correct (we are not including suspect cases in which the outfit is too small). So, we want the probability,

$$\Pr(\text{outfit} \geq 1.3|\text{RM}) = \Pr\left(\chi^2 \geq 30 \times 1.3|\text{RM}\right)$$
$$= \Pr\left(\chi^2 \geq 39|\text{df} = 30\right) = 12.6\%$$

We would conclude that we do not have cause to reject the null hypothesis at 5% significance.

Figure 10 shows probabilities as a function of outfit measure for samples of size $n = 30$ and $n = 300$. If the sample size is $n = 300$, then the probability of an outfit measure greater than 1.3 drops significantly,

$$\Pr(\text{outfit} \geq 1.3|\text{RM}) = \Pr\left(\chi^2 \geq 390|\text{df} = 300\right) = 0.0354\%$$

A p value of 5% is achieved with an outfit value of 1.459.

With a sample of size $n = 2,400$, even an outfit score of 1.1 has a very low p value,

$$\Pr(\text{outfit} \geq 1.1|\text{RM}) = \Pr\left(\chi^2 \geq 2,640|\text{df} = 2,400\right) = 0.0383\%$$

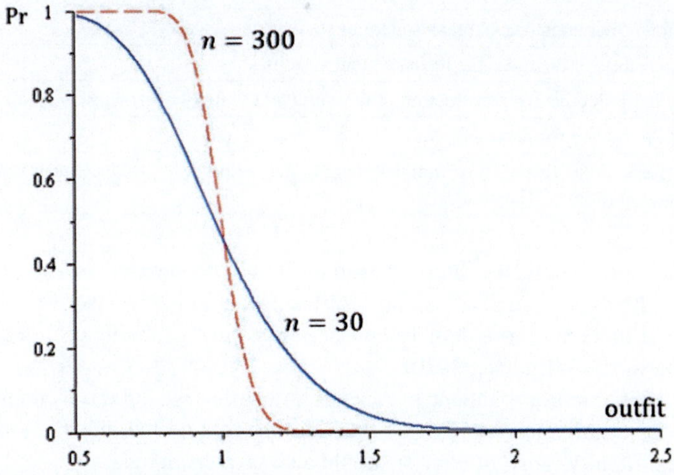

Fig. 10 Probability outfit greater than statistic

Appealing to Wilson and Hilferty (1931), we often find that with large samples, 'the Z-values associated with the infit/outfit values [can be] huge (over 4 to 9.9)' (Linacre, 2020). However, according to Linacre, 2020 https://www.winsteps.com/winman/misfitdiagnosis.htm), a sample size of 2,400:

> *... indicates that even a mean-square of 1.2 (and perhaps 1.1) would be reported as misfitting highly significantly. So [the] mean-squares tell us: "these data fit the Rasch model usefully," and the Z-values tell us: "but not exactly." This situation is often encountered in situations where we know, in advance, that the null hypothesis will be rejected. The Rasch model is a theoretical ideal. Empirical observations never fit the ideal of the Rasch model if we have enough of them*

The outfit value then is a guide to which items in a test are likely to be following the Rasch model. However, outfit is very sensitive to outliers. Infit resolves this problem and is a very useful supplementary statistic for the purposes of calibration. However, outfit is a conventional chi-square, familiar to most statisticians. It is the preferred way to report fitness unless there is a strong reason for reporting infit (Linacre, 2020).

Note that another way to deal with clear outliers is to 'clean up' the data by changing outliers to expected responses. This of course should only be done in the process of calibration. It should not be used as an indicator of model fitness.

Rules of thumb for using infit and outfit seem to follow logical lines. For example, outfit, being a statistical measure with an accompanying population model (at least a widely known and relatively simple model), is useful for reporting and should be the preferred measure for calibration. However, a large difference between outfit and infit would seem to suggest that an outlier is distorting the outfit measure. This should be investigated.

As a rule of thumb, what kind of values should we look for in fitness? Linacre (2020) recommends that values be interpreted as in the following table.

Interpretation of parameter-level mean square fit statistics:	
> 2.0	Distorts or degrades the measurement system
1.5 – 2.0	Unproductive for construction of measurement, but not degrading
0.5 – 1.5	Productive for measurement
< 0.5	Less productive for measurement, but not degrading. May produce misleadingly good reliabilities and separations

As shown previously, the interpretation of the values depends on sample size. If the test involves less than 30 observations, it is probably too insensitive, i.e. 'everything fits'. If there are more than 300 observations, it is probably too sensitive, i.e. 'everything misfits' (Linacre, 2020).

Fitness is an important statistic in the calibration of a test, but it is by no means the only one. Discrimination has already been discussed, and the information balance across the difficulty range of the test should also be considered.

Conclusion

In the text, the affect and significance of sample size have been discussed. In regression generally, the null hypothesis is the result we would like to strongly assert, and sample size often has the reverse impact than on, say, simple tests of the mean. Basic elements of Rasch modelling have also been discussed. 'Tests' using the Rasch model are used to measure latent traits, while fitness is used in the development of the 'test'. Such factors as uniform DIF and information have been examined. It has been shown how the two most-used fitness measures, infit and outfit, are derived from the χ^2 statistic and how they relate to the Rasch model.

Appendix

Dichotomous Model as a Function of Outcome, *x*

Recall that with a dichotomous RM,

$$P(X_{ni} = 1) = \frac{e^{\beta_n - \delta_i}}{1 + e^{\beta_n - \delta_i}}$$

and

$$P(X_{ni} = 0) = 1 - P(X_{ni} = 1) = \frac{1 + e^{\beta_n - \delta_i}}{1 + e^{\beta_n - \delta_i}} - \frac{e^{\beta_n - \delta_i}}{1 + e^{\beta_n - \delta_i}} = \frac{1}{1 + e^{\beta_n - \delta_i}}$$

However, if we let x_{ni} be one of two values, 0 or 1, with a bit of algebra, we can combine these into the expression,

$$P(X_{ni} = x_{ni}) = \frac{\exp(x_{ni}(\beta_n - \delta_i))}{1 + \exp(\beta_n - \delta_i)} \tag{6}$$

Proving this is simply a matter of examining the two possible cases, $x_{ni} = 1$ and $x_{ni} = 0$. Firstly, consider the case $x_{ni} = 1$. We have

$$\begin{aligned} P(X_{ni} = x_{ni}) = P(X_{ni} = 1) &= \frac{\exp(\beta_n - \delta_i)}{1 + \exp(\beta_n - \delta_i)} \\ &= \frac{\exp(1 \times (\beta_n - \delta_i))}{1 + \exp(\beta_n - \delta_i)} = \frac{\exp(x_{ni}(\beta_n - \delta_i))}{1 + \exp(\beta_n - \delta_i)} \end{aligned}$$

In the case that $x_{ni} = 0$, we have the same result (albeit through a somewhat creative route),

$$\begin{aligned}P(X_{ni} = x_{ni}) &= P(X_{ni} = 0)\\ &= \frac{1}{1 + \exp(\beta_n - \delta_i)}\\ &= \frac{\exp(0)}{1 + \exp(\beta_n - \delta_i)} = \frac{\exp(0 \times (\beta_n - \delta_i))}{1 + \exp(\beta_n - \delta_i)}\\ &= \frac{\exp(x_{ni}(\beta_n - \delta_i))}{1 + \exp(\beta_n - \delta_i)}\end{aligned}$$

Thus, we have shown that Eq. (6) holds.
We also have

$$P(X_{ni} = x_{ni}) = \frac{\exp(x_{ni}(\beta_n - \delta_i))}{\sum_{l=0}^{1} \exp(l(\beta_n - \delta_i))} \tag{7}$$

This is easily shown using Eq. (6),

$$\begin{aligned}P(X_{ni} = x_{ni}) &= \frac{\exp(x_{ni}(\beta_n - \delta_i))}{1 + \exp(\beta_n - \delta_i)}\\ &= \frac{\exp(x_{ni}(\beta_n - \delta_i))}{\exp(0(\beta_n - \delta_i)) + \exp(1(\beta_n - \delta_i))}\\ &= \frac{\exp(x_{ni}(\beta_n - \delta_i))}{\sum_{l=0}^{1} \exp(l(\beta_n - \delta_i))}\end{aligned}$$

Let us perform the same analysis with the PL2 model. For this model, put

$$P(X = 1) = \frac{e^{a(\theta - b)}}{1 + e^{a(\theta - b)}} = \frac{1}{1 + e^{-a(\theta - b)}}$$

Here, θ is the person ability, b is the item difficulty and a is the item discrimination. We want to show first

$$P(X = x) = \frac{e^{ax(\theta - b)}}{1 + e^{a(\theta - b)}} = \frac{e^{ax(\theta - b)}}{\sum_{l=0}^{1} e^{al(\theta - b)}} \tag{8}$$

In the case $x = 1$, the result is obvious. Take the case $x = 0$.

$$\begin{aligned}P(X = 0) = 1 - P(X = 1) &= \frac{1}{1 + e^{a(\theta - b)}}\\ &= \frac{e^{0}}{1 + e^{a(\theta - b)}} = \frac{e^{ax(\theta - b)}}{1 + e^{a(\theta - b)}} = \frac{e^{ax(\theta - b)}}{\sum_{l=0}^{1} e^{al(\theta - b)}}\end{aligned}$$

Thus, we have the desired result of Eq. (8).

R Code for Toy Example

```
> #  John Michael Linacre's toy example
> #  Rasch model estimation: Calculating calibrations and mean-squares with JMLE.
> #  3 parts:
> #  https://www.youtube.com/watch?v=LvE8npeSjZ0   3 Feb 2017
> #  https://www.youtube.com/watch?v=X2zaib5VDnk
> #  https://www.youtube.com/watch?v=O6xH2lKbWgc
> #  Code by author
> ############
> # Data
> Mary <- c(1,1,1,0,0)   # Observed results for Mary's test items
> George <- c(0,1,1,0,1) # Observed results for George
> Delta <-c(1,2,3,4,5)   # Item difficulty. This is unchanged throughout this example
> L <- 5 # Number of items
>
>
>
>
> options(digits=3)  # To make the output easier to read (default is 7, max is 22)
>
> # Model
> df <- data.frame(Mary,George); t(df) # Transpose to display data in same form as in the video
       [,1] [,2] [,3] [,4] [,5]
Mary      1    1    1    0    0
George    0    1    1    0    1
> t(Delta)
     [,1] [,2] [,3] [,4] [,5]
[1,]    1    2    3    4    5
>
> # Probability of success from logit difference
> # All based on Delta
> # Note that probability of FAILURE is 1-Pr
> Pr <- function(theta){
+    e <- exp(theta - Delta)
+    P <- e/(1+e)
+    return(P)
+ }
>
> # Variance calculation
> variance <- function(theta){
+    Prob <- Pr(theta)
+    return(Prob*(1-Prob))
+ }
>
> #### Observed raw (total) scores
> observed <- sum(Mary); observed
[1] 3
> sum(George)  # Both have the same observed total score
[1] 3
>
> #### If the ability was 3 the chances of success on the different items are
> Pr(3)
[1] 0.881 0.731 0.500 0.269 0.119
> Pr(3.5)  # Chances if ability was 3.5
[1] 0.924 0.818 0.622 0.378 0.182
>
```

```
> #### The expected total score with ability of 3 is
> sum(Pr(3))
[1] 2.5
> sum(Pr(3.5))  # Expected score if ability is 3.4
[1] 2.92
> # No we are getting close
>
> # Bernouli variance for each question
> # Also called statistical information (Fisher?)
> variance(3)
[1] 0.105 0.197 0.250 0.197 0.105
> variance(3.5)
[1] 0.0701 0.1491 0.2350 0.2350 0.1491
>
> # Total variance
> Tvar <- sum(variance(3.5)); Tvar
[1] 0.838
>
> # Next apply Newton-Raphson (N-R) method
> # Part 1 @ 19:00
> # Applying it to observed score = expected total score
> # B' <- B + (observed-expected)/total variance
> B <- 3.5; observed <- sum(Mary); expected <- sum(Pr(B)); expected
[1] 2.92
> # One step of N-R
> B <- B + (observed-expected)/Tvar; B  # New estimate
[1] 3.59
>
> # One step of N-R defined as a function
> NR <- function(B){
+    expected <- sum(Pr(B))
+    B <- B + (observed-expected)/Tvar
+    return(B)
+ }
>
> # Rapid convergence of N-R
> options(digits=22)  # We don't need this level of precision but it is interesting
> B <- 3
> B <- NR(B); B # Next iteration
[1] 3.5963711684465274
> B <- NR(B); B
[1] 3.5908301631430106
> B <- NR(B); B
[1] 3.5907898708164177
> B <- NR(B); B
[1] 3.5907895873056281
> B <- NR(B); B
[1] 3.590789585311231
> B <- NR(B); B
[1] 3.5907895852972018
> B <- NR(B); B
[1] 3.5907895852971028
> B <- NR(B); B
[1] 3.5907895852971023
> B <- NR(B); B
[1] 3.5907895852971023
> options(digits=3)
> # Let us save the optimal value
> BM <- B
```

```
>
> # Notice that the same works for George, because the estimate is based on the raw/observed score
> # But what if we do take into account the actual separate scores?
> # Part 1 - 21:40 (The way that Ronald Fisher did it)
>
> # He was looking for the most likely estimate
> # Let's look at the probabilities of getting the response for the individual
> # To start with, suppose B=1
> B <- 1; Pr(B) # Probability of success
[1] 0.5000 0.2689 0.1192 0.0474 0.0180
> 1-Pr(B)     # Probability of failure
[1] 0.500 0.731 0.881 0.953 0.982
>
> # How can we show just the probabilities for Mary?
> # Firstly, to find the probabilities for Mary successes, just "multiply" Mary times the proba-bilities of success
> Mary * Pr(B)  # We need to discount the zero entries
[1] 0.500 0.269 0.119 0.000 0.000
> # We can use the trick that Mary's failures are:
> 1-Mary
[1] 0 0 0 1 1
> # The probabilities for Mary's failures are then 1-Mary times the probabilities of failures
> (1-Mary)*(1-Pr(B))
[1] 0.000 0.000 0.000 0.953 0.982
> # If we add the successes and fails we have the overall probabilities for Mary
> Mary*Pr(B)  +  (1-Mary)*(1-Pr(B))
[1] 0.500 0.269 0.119 0.953 0.982
>
> # Turn this into a function giving:
> # Probability of RESULT for person given ability (theta)
> P <- function(theta, person){
+    Prob <- Pr(theta)
+    s <- Prob*person  # successes
+    f <- (1-Prob)*(1-person) # fails
+    return(s+f)
+ }
>
> P(B, Mary)  # Check
[1] 0.500 0.269 0.119 0.953 0.982
>
> # Multiply them together to give the likelihood
> prod(P(B, Mary)) # Likelihood
[1] 0.015
>
> # Let's make this a function:
> Lambda <- function(theta, person){
+    return(prod(P(theta,person)))
+ }
>
> Lambda(1, Mary)  # Check
[1] 0.015
> Lambda(2, Mary)  # What if B=2?
[1] 0.0825
> Lambda(3, Mary)  # What if B=3?
[1] 0.207
> Lambda(4, Mary)  # What if B=3?
[1] 0.224
> Lambda(5, Mary)  # What if B=3?
[1] 0.111
> Lambda(BM, Mary)  # What if B is the optimal value?
```

```
[1] 0.24
>
> # We can plot likelihood against ability estimate
> # Unfortunately, we need to separate out each x value in order to plot the values
> fPoints <- function(theta, person){
+   y <- c()
+   for(x in theta){  # To take each theta value separately
+     y <- c(y,Lambda(x,person))
+   }
+   return(y)
+ }
> B <- 1:6; B
[1] 1 2 3 4 5 6
> fPoints(B,Mary); fPoints(B,George)
[1] 0.0150 0.0825 0.2073 0.2242 0.1108 0.0298
[1] 0.000275 0.001511 0.003797 0.004106 0.002029 0.000546
>
> #  Plot the possible values for beta as a function of mu
> Lo <- 1; Hi <- 6  # Set low and high x-axis plot limits
> B <- seq(1,6,0.02)
> plot(B,fPoints(B,Mary),ylim=c(0,0.25), xlab="Ability", ylab="Likelihood",type="l") # draw a graph
> lines(B,fPoints(B,George),lty=1,col="red") # draw a graph
> # We see that George's likelihood are much, much lower
>
>
> # Also, the two methods give the same result
> # The likelihood curves are simply scaled versions of each other
> # Even though George's result is much less expected than Mary's
> B <- 1:6; B
[1] 1 2 3 4 5 6
> fPoints(B,Mary)/fPoints(B,George)
[1] 54.6 54.6 54.6 54.6 54.6 54.6
> # Only the raw score determines the optimum (in Rasch)
> # Part 2 @ 12:00
>
> # Note that log-likelihood is used in a computer, and sometimes in theory
>
> # TAKE A BREAK Part 2 13:00
>
> # When everything is nice and well behaved
> # When Difficulties are known, we just have one likelihood curve
> # We can always check with points on either side of the peak
>
> # But with Joint Maximum Likelihood Estimation (JMLE),
> # also called Unconditional maximum likelihood estimation.
> # This leads to a nonconvex surface (see "Nonconvex and Convex Functions of")
> # Iterative process can hop around a minor maximum, or can hop between minor maximums
> # Missing data exacerbates this.
> # Part 2 @ ???
> # Instead of using the N-R approach we can switch over to using a curve fitting approach.
> # Curve-fitting approach
> # We use the relationship of a logistic ogive - much more robust
> # Winsteps uses it
> # Estimates with categories with a large disparity in number of items
> >
> ############################
> #  FITNESS
> # Part 2 @ 22:00 "Now let's think about the fit side"
> ############################
```

```
> # Mary has expected set of responses
> # Gerge has an UNexpected set of responses
>
> # There are many fit statistics that have been defined for this type of data
> # We look at two of them, outfit, related to standard chi-square measure and
> # infit, which is an information weighted form of statistic
> # One term of the chi-square measure:
> #                    (observed - expected)^2/variance
> # This has an expected value of one.
> # If exact agreement, then a value of zero. This is impossible in this case.
> # We are predicting "too well" Part 2 @ 26:50
>
> # Take expected values for George, for example
> George                        # Observed
[1] 0 1 1 0 1
> Pr(BM)            # Expected
[1] 0.930 0.831 0.644 0.399 0.196
> v <- variance(BM); v          # Variance
[1] 0.0649 0.1406 0.2294 0.2398 0.1578
> Out <- (George - Pr(BM))^2/v; Out      # Individual chi-square values
[1] 13.340  0.204  0.554  0.664  4.093
> sum(Out)
[1] 18.9
> outG <- sum(Out)/L; outG  # Outfit for George
[1] 3.77
>
> # Take expected values for Mary,
> # The only thing that changes are the data values
> Mary                          # Observed
[1] 1 1 1 0 0
> Out <- (Mary - Pr(BM))^2/v; Out        # Individual chi-square values
[1] 0.075 0.204 0.554 0.664 0.244
> sum(Out)
[1] 1.74
> outM <- sum(Out)/L; outM  # Outfit for Mary
[1] 0.348
>
> # Mary's outfit value is much lower than George's
> # In fact, Mary's response follows the Guttman pattern
> # Succeeding on everything easy and failing on everything hard
> # Is there some reason for this  Part 3 - 6:00
> # Something is constraining the data
>
> # INFIT
> # Part 3 - 8:00
> # If there is one unexpected response it adds a lot to the outfit measure
> # For example, George's first response should be correct,
> # but the individual chi-square value is 13.340
> # Outfit is very sensitive to individual, unexpected responses.
> # The Infit statistic, also called the information weighted statistic,
> # adds up all the tops and divide by all the bottoms
> InG <- sum((George - Pr(BM))^2)/sum(v); InG   # Individual chi-square values
[1] 2.19
> InM <- sum((Mary - Pr(BM))^2)/sum(v); InM     # Individual chi-square values
[1] 0.431
```

References

Andrich, D. (2004). Controversy and the rasch model: A characteristic of incompatible paradigms? *Medical Care, 42*(1), I7–I16.

Bond, T. G., & Fox, C. M. (2001). *Applying the rasch model: Fundamental measurement in the human sciences.* Routledge.

Hatzinger, R. (2010). *Rasch models and the R package eRM.* https://eeecon.uibk.ac.at/psychoco/2010/slides/Hatzinger.pdf, October 9, 2018.

Linacre, J. M. (2001). *Rasch model estimation: Calculating calibrations and mean-squares with JMLE. Linacre, 2001.* https://www.youtube.com/watch?v=LvE8npeSjZ0. Jun 28, 2020.

Linacre, J. M. (2020). *Fit diagnosis: infit outfit mean-square standardized,* Winsteps(R). https://www.winsteps.com/winman/misfitdiagnosis.htm. June 22, 2020.

Magis, D., Yan, D., & von Davier, A. A. (2017). *Computerized adaptive and multistage testing with R: Using packages catR and mstR.* Springer Publishing Company.

Nuzzo, R. (2014). Scientific method: Statistical errors. *Nature, 506*(7487), 150.

Smith, A. B., Rush, R., Fallowfield, L. J., Velikova, G., & Sharpe, M. (2008). Rasch fit statistics and sample size considerations for polytomous data. *BMC Medical Research Methodology, 8*(33), 33.

Walker, A. A., Jennings, J. K., & Engelhard, G. (2018). Using person response functions to investigate areas of person misfit related to item characteristics. *Educational Assessment, 23*(1), 47–68.

Wikipedia. (2021). *Rasch model,* Wikipedia. https://en.wikipedia.org/wiki/Rasch_model, Jul 2, 2021.

Wilson, E. B., & Hilferty, M. M. (1931). The distribution of Chi-square. *Proceedings of the National Academy of Sciences of the United States of America, 17*(12), 684–688.

Wright, B., & Panchapakesan, N. (1969). A procedure for sample-free item analysis. *Educational and Psychological Measurement, 29*(1), 23–48.

Wu, M., & Adams, R. (2007). *Applying the Rasch model to psycho-social measurement: A practical approach.* Educational Measurement Solutions.

Ian Storey Ph.D., is Lecturer and Researcher in Torrens University in Melbourne. His research interests include information security and risk management, particularly accountability, as well as EA architecture, control stabilization and e-learning. He has lectured in information security and risk management over an extensive period of time and has liaised with industry partners in that time. His Ph.D. was on the topic of genetic algorithms and multi-objective optimization.

Data Evaluation of Happiness Scale Online Study: A Rasch Measurement Analysis

Elspeth McKay, Keven Asquith, Eugenia Smyrnova-Trybulska, Anna Porczyńska-Ciszewska, and Tomasz Kopczyński

Abstract Some say the rise of robots and automation are predicted to lead to the loss of 20 million manufacturing jobs by 2030 (Redd and Klonoski in How Robots Change the World: What automation really means for jobs and productivity, Oxford Economics, June. How Robots Change the World—What automation really means for jobs, productivity and regions, 2019). As a result, the education sector is facing massive restructuring to manage this predicted visionary future. Researchers in Poland, examining a happiness scale key (Skala Doświadczania Szczęścia—SDS), found variance in happiness levels of students affected their capacity to undertake robotics programming. The researchers were keen to further ratify their findings through Rasch measurement theory (RMT) modelling. This work carried out independently by data analysts has provided a fine-grained interpretation of the correlation between happiness levels and Polish classroom robotics pedagogies.

Introduction

As the demand for peoples' social needs using robots instead of human beings grows, researchers seek guidance for ethical and due diligence concerning people's health

E. McKay (✉)
Cogniware, Melbourne, Australia
e-mail: e.mckay@cogniware.com.au

K. Asquith
3D-Geo Pty Ltd, Level-5, 8-Market Street, Melbourne, Australia
e-mail: keven@3d-geo.com

E. Smyrnova-Trybulska · A. Porczyńska-Ciszewska · T. Kopczyński
University of Silesia, Bankowa 12, 40-007 Katowice, Poland
e-mail: esmyrnova@us.edu.pl

A. Porczyńska-Ciszewska
e-mail: anna.porczynska-ciszewska@us.edu.pl

T. Kopczyński
e-mail: tomasz.kopczynski@us.edu.pl

E. McKay (ed.), *Manage Your Own Learning Analytics*, Smart Innovation, Systems and Technologies 261, https://doi.org/10.1007/978-3-030-86316-6_4

and welfare (Guglielmelli, 2014). Even in 2015, emerging devices ranged from drones to humanoids which can play poker, author cookbooks and engage people in conversations (Devlin, 2015). Today, industrial robots and integrated automation of production are in demand in many social activity areas: general and military industries; medicine; aircraft; space industry (self-propelled vehicles based on robotic systems); service sector; and domestic life. Robots change the way we live and work. This situation also means that companies urgently need specialists to design, construct and program robots (Morze et al., 2018). Moreover, the digital disruption caused by the COVID-19 pandemic has quickened the adoption pace at a dazzling rate and is hard to keep up with; one such vehicle for managing this information is to know where the industry sector is heading (Crowe, 2021).

The education sector has also followed this trend by offering courses to equip students with advanced digital technical skills required to enter this dramatic shift in the social need aspects of digital transformation. For example, a recent study in Poland examined whether there was any correlation between using robotics programming in classroom tutorials and general happiness levels. The primary aim of this paper is to report on the data analysis undertaken by that study.

Background

The current global pandemic related to the SARS-CoV-2 virus has undoubtedly introduced significant changes in Polish lifestyles (and not only) in education, somehow intensifying students' contact with modern information technologies (IT), impacting the happiness and mental well-being of young people. While it is safe to assume that every human being has a desire to be happy, there is no universal definition of this pleasant emotion (Pogosyan, 2016). Research shows that different cultures worldwide have unique concepts of happiness and well-being (Englemann & Pogosyan, 2013).

Undoubtedly, people's relationship with the digital age tools is of interest to many scientific disciplines. A relatively new scientific exploration area that integrates many different disciplines is human–computer interaction (HCI), dealing with the appearance, evaluation and implementation of interactive computer-based systems used by people and exploring the phenomena related to this area (Plichta, 2017).

The primary purpose of this chapter is to report the empirical data evaluation of the happiness scale of an online research study conducted in Poland, using the Rasch measurement theory (RMT) data analysis model. This work extends an earlier factor analysis conducted in Poland, distinguishing two factors, explaining 38% of the measurement results variance.

The first factor defined the sense of commitment and satisfaction accompanying the achievement of one's own goals. The research showed this factor explained 20% of the variance. The content of the questions about the highest load of the discussed aspect indicated experiencing happiness (in the meaning of the concept of flow experience) or being in the zone; meaning the mental state of someone wholly immersed in an activity most often accompanies a person during the total commitment to consciously selected goals and aspirations and the ways of their implementation (Hill, 2016). This factor also explained the relationship between experiencing happiness and positive emotional states, such as experiencing satisfaction, well-being and mood, feelings of elation and joy, and deep contentment.

The second factor that explained 18% of the variance was called active stimulation of self-well-being. Experiencing happiness seemed to be a consequence of one's activity, i.e. taking active measures for mental well-being. The questions that were to the highest degree saturated with this factor concerned the individual's involvement in various tasks, taking up challenges and enjoying one's good mental condition.

Literature Review

Through the lens of experiencing happiness in the context of conducting classes with the use of modern IT pedagogies, a review was conducted of social robots in classrooms in emerging educational technology and teacher education between 2008 and 2009, by Rosanda and Starčič (2019), which revealed robotics applied in classrooms in different subjects and academic levels. Morze et al. (2018) p: 21 emphasised that.

> *Robotics is an applied science about the design, development, operation, and use of robots. Nowadays, increased attention is paid to robotics as to applied science, including its educational and developmental potential. This has created a new trend in education: educational robotics. Educational robotics is a cross disciplinary area of students' learning. Its process integrates the knowledge of STEM subjects (physics, technology, mathematics), cybernetics, mechatronics, and informatics.*

Yang et al. (2020) described an engaging educational experience concerning artificial intelligence (AI) robotics-based upon voice interaction realised personalised pedagogies. Moreover, these researchers from Korea and China concluded that such AI-based robotics for physical education pedagogies helped improve classroom efficiencies of the physical education teaching and improved student interest. Furthermore, the literature analyses proved that the classes with robotics could positively influence learners' different competencies in various ages—from earlier childhood education to students of their first and second degrees, and adult specialists (Yang et al., 2020).

Jiménez et al. (2020) described a two-wheeled auto-balancing robot's modelling and control as an exciting didactic platform for control engineering education. The authors analysed the system response's design against the two implemented control algorithms simulated using the commands and toolboxes available from MATLAB/Simulink. According to Jiménez et al. (2020), this work expected to provide the basis for further development and research.

A study conducted by Smyrnova-Trybulska et al. (2016) researched primary school students' opinion of their teachers using educational robots in their classrooms. The main focus of this study was the data analysis of the students' view on

STEM education for young learners. In particular, they concluded that "*classes in robotics—properly taught—will have an impact on the development of mathematical literacy and scientific-technical information and social competencies*" (p.197). However, the authors also emphasised that "*Simultaneously, still open is the question concerning the comprehensive STEM education of prospective teachers pursuing pedagogical programmes, particularly in the specialisation of early education*" (p. 204).

Compared to eastern countries, learning environments of Western countries vary on critical pedagogical differences between studies of robotics undertaken in the West versus the East. For instance, Yang et al. (2020) in China and Korea concentrate on the machine-dimension of the HCI (McKay, 2008) concerning the mechanical robotic control design, while Crowe (2021) in the USA and Devlin (2015) in Britain focus on the human-dimension or social HCI aspects of robotics (McKay, 2008). However, Jiménez et al. (2020) in the Republic of Columbia reflect the Eastern preference to report their research through the lens of the digital HCI design requirements, ignoring the human-dimension).

Research Design and Methodology

The research design of the happiness scale key 'skala doświadczania szczęścia' (SDS)—in Polish, comprised an online questionnaire for young people aged 15–20 years old. The main goal was to examine their experience and satisfaction in robotics classes and any relationship to their general happiness levels. This research captured anonymous student responses to a three-part online questionnaire anticipated to take 10–15 min to complete.

Part: I involved capturing the timestamp (date and time), participant profile details, age, high school grade, gender, living location and type of school.

Part: II involved a survey on participating in classes with robotics programming and educational kits (25-questionnaire items) that involved an initial question regarding whether the participant had any activities with robotics programming and educational kits; they were to respond with a yes/no answer. In the case of yes, participants answered the following questions and rated their response to their satisfaction level on a scale from 1 to 7, where 1 meant complete dissatisfaction and 7 indicated the maximum level of satisfaction. Part: II asked students how their experience with robotics programming helped develop knowledge and various technical/social skills and how often they would like to have robotics classes.

Part: III was a survey on the level of happiness experiences (40-questionnaire items). These items contained a series of statements with which people usually describe themselves as concerned with student experiences and feelings in various life situations. Part: III commenced with the following statement regarding the scale providing information about the frequency of their happiness and the accompanying circumstances. Because every person is different, there were no right and wrong answers, only answers that said something about the participant undergoing the questionnaire. This study's results were for scientific purposes only, so the answers participants needed to give were truthful. The researchers guaranteed anonymity to all respondents.

Participants were to read the 40-Part: III statements carefully and determine how they applied to them and their inner experiences. The questionnaire advised students not to spend too much time on individual statements and not go back to previous answers. In addition, the questionnaire had a 7-point Likert scale to express their happiness feelings to determine how often the following sensations accompanied them by marking a rating score selected from this scale: 1-never; 2-very rarely; 3-rarely; 4-sometimes; 5-often; 6-very often; and 7-always.

Research Questions

The research investigated the relationship between the experience of happiness and high school students' participation in robotics programming and educational kits.

The research problem, which is the subject of the undertaken research, took the form of the following research questions:

1. Are students who work with robotics programming and educational kits happier than those who do not work with robots?
2. Is there a relationship between experiencing happiness and high school students' assessment of robotics programming as interesting and cognitively developing?
3. Is there a connection between experiencing happiness and high school students' assessment of robotics programming classes to create knowledge in other subjects?
4. Is there a relationship between experiencing happiness and evaluating robotics programming classes to develop creativity and create high school students' projects?
5. Is there a relationship between experiencing happiness and evaluating robotics programming classes to develop high school students' teamwork?
6. Is there a relationship between experiencing happiness and evaluating robotics programming classes to develop social competencies?
7. Is there a relationship between experiencing happiness and evaluating robotics programming classes to develop communication competencies?
8. Is there a relationship between the number of hours of classes with robotics programming and the feeling of happiness?
9. Is there a relationship between the prospect of a future job or career in robotics training and happiness?
10. Is there a relationship between satisfaction from robotics programming classes, happiness, and the teacher's competencies?

Results

The results of the questionnaire were analysed using the RUMM2030 Prof software application. This unidimensional psychometric Rasch measurement tool developed at the University of Western Australia (RUMM 2030, 2015) presents an opportunity for conjoint measurement of participant ability and latent trait (knowledge constructs) on the same linear scale divided into equal intervals along the scale (Andrich, 2011). An essential feature of a Rasch analysis is to ensure that the data fit the model. Other measurement models follow the item response theory (IRT), where researchers search for a model to fit the data. See Fig. 1 for a visual representation of the data analysts evaluation plan.

Data Preparation

The first step of data analysis is a manual (or visual) check of the data file (Fig. 2). The left-hand table below shows a section of the original data file with 128-participants (identified in the left-hand column as Part: I PersonID data) and the remaining columns representing the participant's Likert scoring answers to the online questionnaire with 65-items (see Appendix for the complete listing). In the data sent by the researchers for this Rasch measurement analysis, the Part: III survey (40-items) was placed before Part: II (25-items). A scan of this data set indicated that some participants did not complete sections of the two surveys, and therefore the data analysis was segmented with extreme records removed. This process is

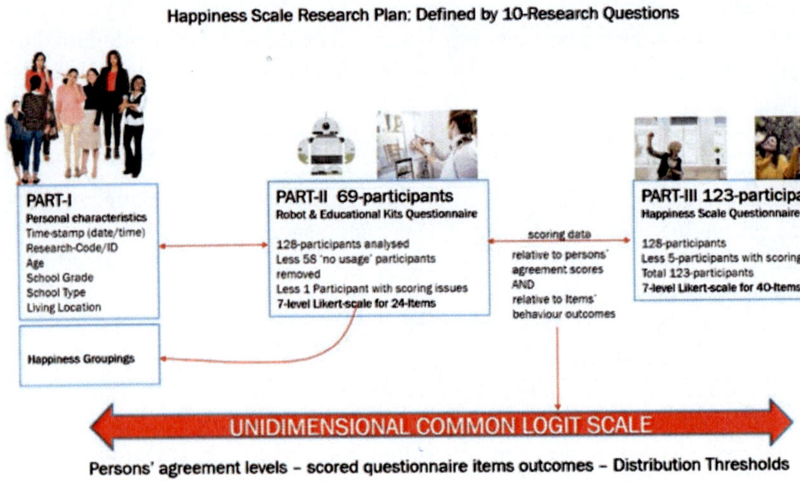

Fig. 1 RUMM2030 summary statistics—Full data set and with extremes removed

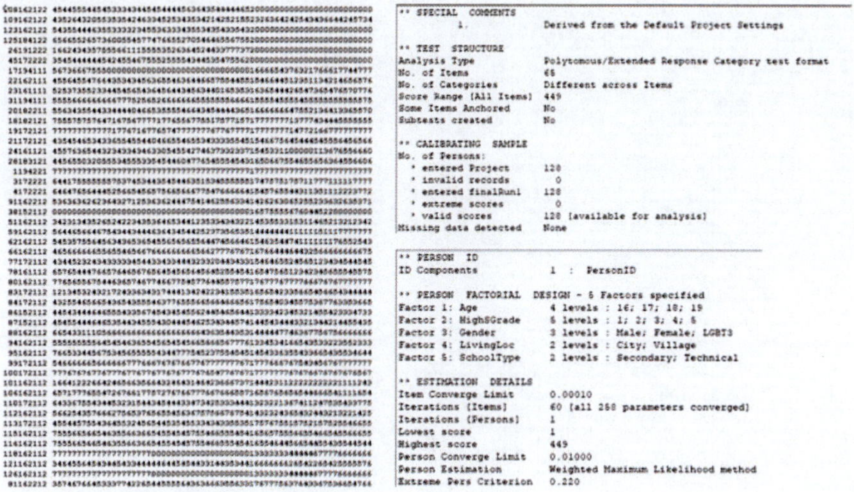

Fig. 2 Coded RUMM2030 data file

discussed fully in the following sections. The analysis needed to determine participants' overall agreement level to the SDS questionnaire items and whether individual item responses had expected outcomes or had anomalous distributions. The right-hand table below displays some of the initial Rasch measurement analysis design, with only 60-iterations required to achieve convergence of all 258-parameters.

However, as noted above, there were anomalous records within this initial data set (Fig. 2), and therefore, the estimate outcome is unreliable. Consequently, the data file was reviewed to determine which records should be removed. The tables below summarise the revised estimate run, with 65-participant records (i.e. with 63-anomalous records removed—mostly participants with no experience with robotics programming and educational kits and therefore no answers for Part: II of the survey). The right-hand table indicates that a reliable outcome was achieved with all 256-parameters converging within 998 iterations (Fig. 3).

Figure 4 shows the summary statistics (unidimensional Rasch logit scale location and fit residual statistics) for the original data on the left and the filtered data on the right to investigate the effect of removing the extreme records. The person data was most affected with the original mean under 0.1 logits and a fit residual mean of almost -1, while the filtered data mean was 0.44 with a very good mean fit residual of only -0.03 (variation between the model curve and actual data values).

A preliminary examination of the questionnaire results in terms of individual participant/person Likert scores relative to each participant's score and each questionnaire items' behaviour, relative to each questionnaire item on a common unidimensional scale, is displayed person-item location distribution below (Fig. 5). The threshold distribution for participants shown in red boxes above the logit scale, whereas the questionnaire/item distribution shown in blue boxes under the logit scale (Fig. 5).

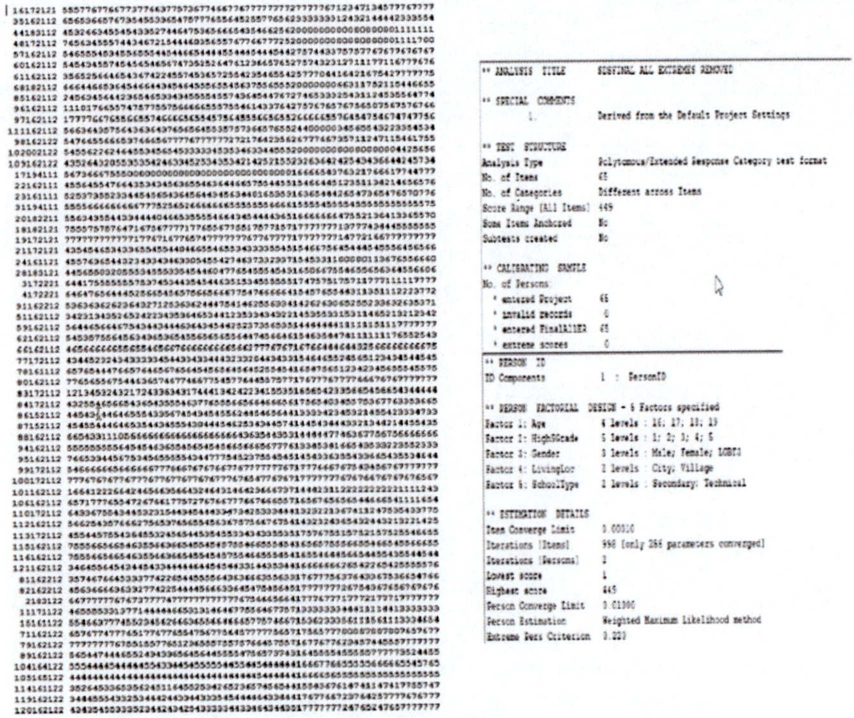

Fig. 3 Coded RUMM2030 data file with 63-anomalous records removed

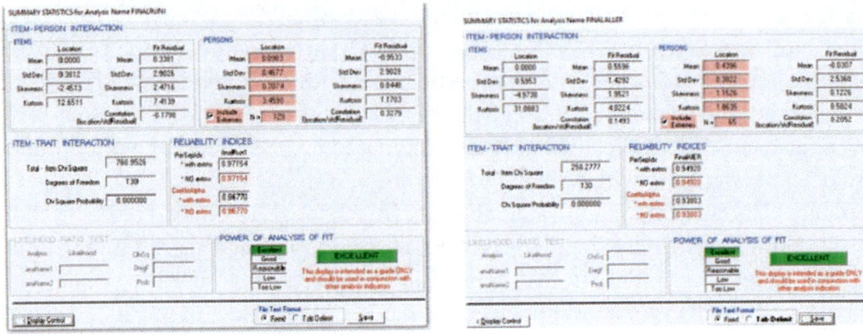

Fig. 4 RUMM2030 summary statistics—full data set and with extremes removed

The original data set had a bimodal distribution of participant scores due to the many participants who did not answer most Part: II questions. Thus, the mean of 0.098 was influenced by all the zero scores for the Part: II questions. The revised data set with extremes removed shows a more normal distribution with a higher mean (0.44 logits) and smaller standard deviation.

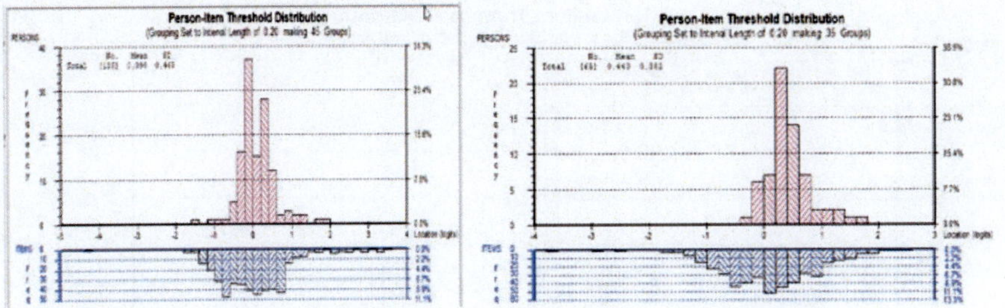

Fig. 5 Person-item location distribution—Full data set and with extremes removed

However, the filtered data set only had 65 of the original 128-participants. Thus, many participants removed would have had valid questionnaire answers for Part: III but were removed because they had no experience with robotics programming or educational kits. As a result, they would have had zeros for the Part: II questions. Conversely, a few participants had robotics experiences in classes but did not answer the Part: III questions to differentiate their answers across the 40-items.

Since the two parts of this experiment are independent (Part: II - robotics programming experience in classes and Part: III - general happiness and satisfaction), the data analysis can separate the two parts, giving a more robust analysis for each individual part.

Part: II Data Review

The Part: II data set would still have participants without classroom experience with robotics programming and educational kits removed. Figure 6 shows that 69-participants remained for analysis with a mean logit score of 0.456 and a standard deviation of 0.581. The distribution is bimodal with a mean fit residual mean of -0.151.

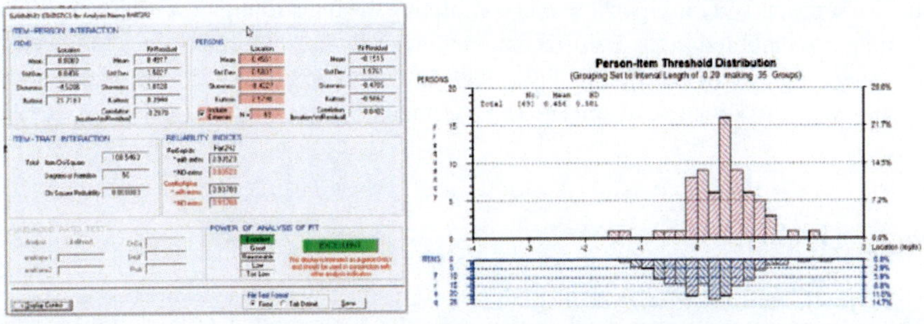

Fig. 6 Part: II analysis summary statistics and person-item location distribution—with extremes removed

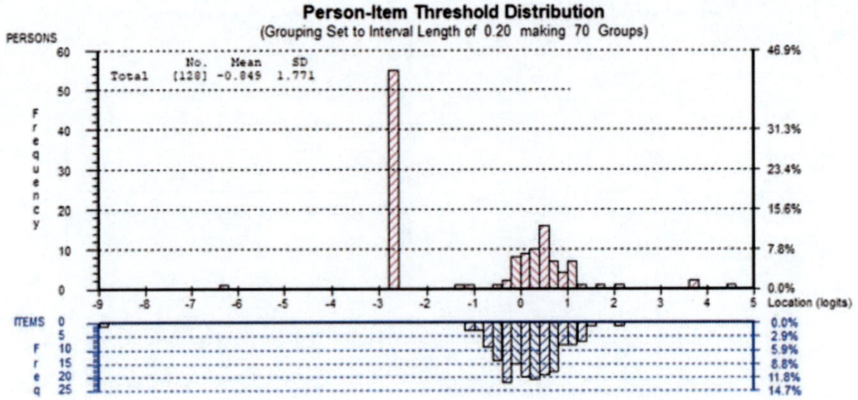

Fig. 7 Part: II analysis person-item location distribution—complete data set

To illustrate the effect that the YES/NO question that starts the Part: II questionnaire and the participant's influence with NO robotics programming experiences in class, and therefore had 0s for the remainder of the questionnaire, the person-item threshold distribution for the full 128-participants is shown in Fig. 4. Almost half the participants have a logit scale value of minus 2.7 (the 'no experience participants'), reducing the mean logit score to -0.849 with a large standard deviation of 1.77. Also, note the first item (YES or NO) located at minus 9-logits compared to the other 24-items scored out of 7. This item was removed from the data analysis. However, this group's analysis was included in the integrated Part: II/Part: III correlation study (Fig. 7).

Part: III Data Review

The Part: III data set was a more robust research design than Part: II, with all 40-items having the 7-level Likert scaling, although three items (13, 18 and 23) had inverse scoring, and most participants completing the questionnaire. There were 123-participants included in the Part: III data review (only five extreme values removed). The mean logit score was 0.508, with a standard deviation of 0.542. The distribution is normal with skewness of only 0.38. The mean fit residual is also low at -0.421 (Fig. 8).

Data Analysis

The data preparation identified issues with the integrated data set, with conditional/optional questions. For example, one dichotomous scored Yes/No question in the Part: II questionnaire corrupted the Rasch model estimates. However, reviewing

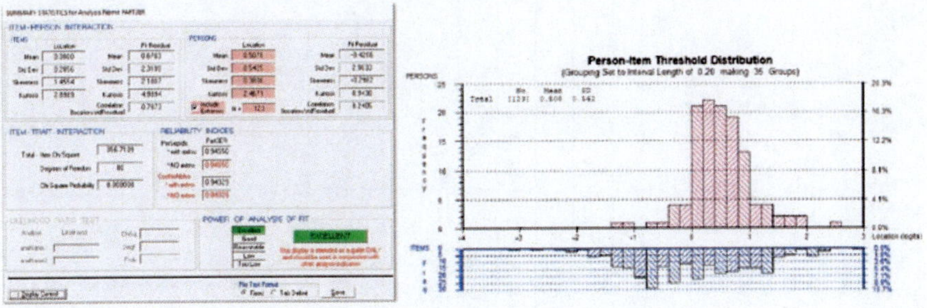

Fig. 8 Part: III analysis summary statistics and person-item location distribution—with extremes removed

the individual Parts (II and III) and removing extreme participants delivered more robust model estimates with good fits to the observed data.

The data analysis focussed on participant happiness levels, the Part: III questionnaire. The data set had five PersonID Part: I variables analysed against the filtered Part: III Rasch outcomes concerning: age; high school grade; gender; living location; and school type. The table below shows the average Part: III logit score for each of the variable categories (Table 1).

This data indicated several interesting trends. First, the age and high school grade variables generally showed an increasing trend in happiness levels from age 15 up to age 18 (0.320 to 0.622) and high school grade 1 to 3 (0.53 to 0.75). However, that trend reversed for 19 year olds (0.38) and high school grades 4 and 5 (0.317 and 0.143, respectively).

There was no significant happiness variation between gender groups, with males, females and the LGBT participant having an average happiness level around 0.515 logits.

The data indicated that village-based participants were generally happier than city-based participants (0.574 compared to 0.463); the secondary school participants were marginally happier than the technical school participants (0.587 compared to 0.488).

Correlation Study

The research project was designed to determine a relationship between participants' exposure and satisfaction level with robotics programming and educational kits and general happiness. The correlation study needed to evaluate the relationship between Part: II (a survey on students participating in robotics programming classes and educational kits) and Part: III (a survey on the level of happiness experiences). The data analysts grouped and analysed the Part: II data and the Part: III Happiness level data of each Part: II group.

Table 1 Part: III Happiness levels

Part: III Happiness levels

Age		High school grade		Gender		Living Location		School Type	
Value	Logit average	Value	Logit average	Value	Logit average	Value	Logit average	Value	Logit average
15	0.320	1	0.530	Male	0.507	City	0.463	Secondary	0.587
16	0.473	2	0.508	Female	0.516	Village	0.574	Technical	0.488
17	0.562	3	0.751	LGBT	0.525				
18	0.622	4	0.317						
19	0.380	5	0.143						

The data groupings developed were based on ten research questions; the following analysis addresses those questions in sequential order.

Research Question-1

Are students who work with robotics/sets happier compared to those who do not work with robotics.

The Part: II questionnaire was a survey about various aspects of robotics usage and satisfaction levels. Therefore, the RUMM2030 logit scale participant distribution from the filtered Part: II data was used to group students. Part: II participants were divided into four groups: those answering NO to having classroom experience with robotics (group-1); those with a low satisfaction level on their robotics experience (a logit score on the filtered Part: II analysis below 0.20 logits, group-2); those with moderate satisfaction with their robotics experiences (between 0.20 and 0.60 logits, group-3); and those having a high satisfaction level with their robotics experience (a logit score above 0.60, group-4). The item characteristic curve (ICC) in Fig. 6 illustrates where these groups plot against the model estimate curve for two questions in the Part: II survey, item-6 and item-16. Note that group-1 was not plotted on these graphs as the large group of students with no experience with robotics have a logit value below minus 3 (Fig. 9).

The table below shows the average happiness level logit value from Part: III for each of the four robot experience groups in Part: II (Table 2).

The data showed an increasing level of overall happiness with increased robotics usage and satisfaction. Participants with no classroom experience with robotics programming or educational kits (group-1) had an average logit score happiness level of only 0.381. Participants with some robotics experience in the classroom but not satisfied with that experience (group-2) had a slightly higher average happiness level at 0.488 logits. Students with moderate satisfaction with their robotics experiences (group-3) were again higher at 0.548. However, participants with a high satisfaction level on their classroom robotics experience (group-4) had an average happiness level of 0.762 logits, which doubled that of the participants with no robotics experience.

Plotting these groups against individual items illustrates how each group compared to the model estimate curve for that model. For item-16 on the Part: III Happiness survey, the high satisfaction group (purple) and the moderate satisfaction group (green) plot above the model estimate and the no usage group (blue) generally plots below the estimated curve. There is considerable variance from item to item and even within one item. However, overall the grouping hierarchy is maintained on the Part: III Items (Fig. 10).

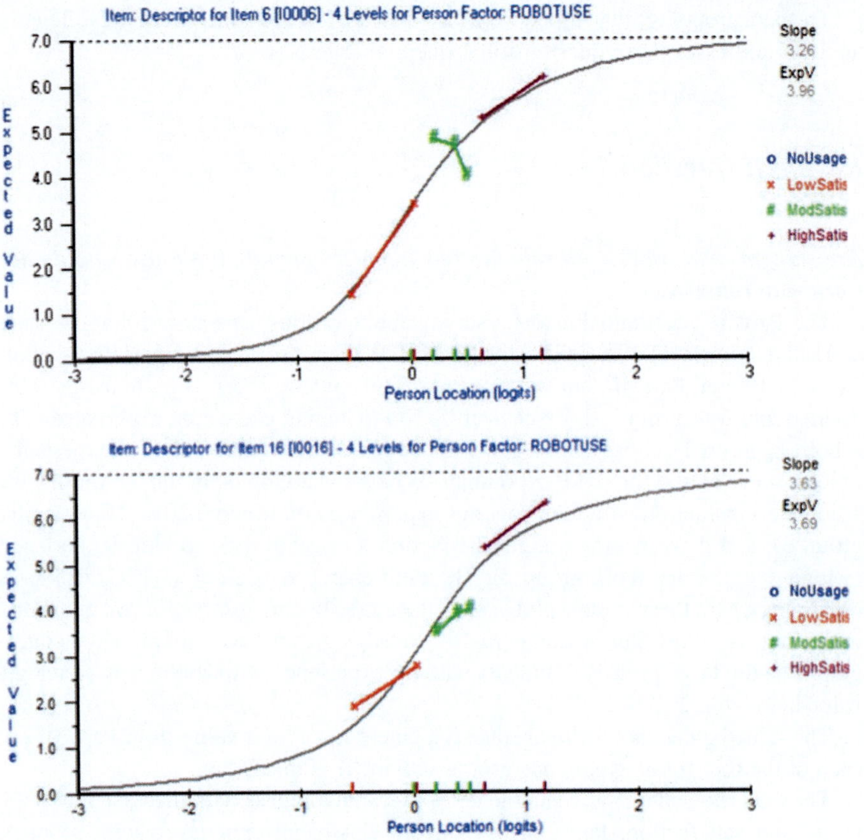

Fig. 9 ICC for Part: II items-6 and 16, with research question-1 groupings

Table 2 Part: III Happiness levels—robotics usage

Part: III Happiness levels	
RQ 1: Robot usage/satisfaction	
Value	Logit average
No usage	0.381
Low satisfaction	0.488
Mod satisfaction	0.548
High satisfaction	0.762

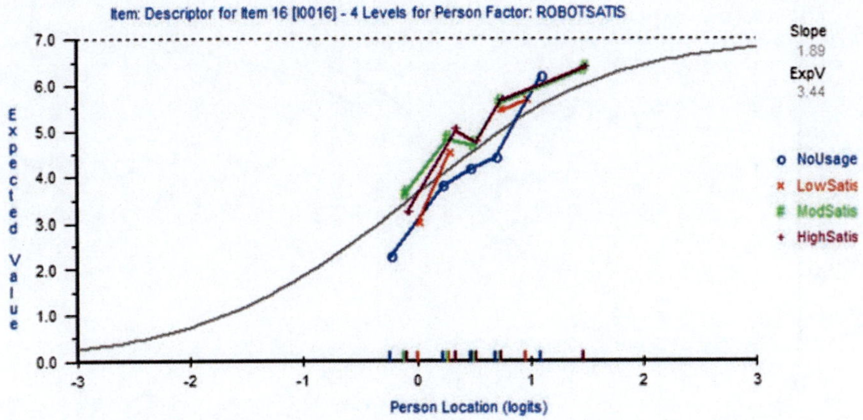

Fig. 10 ICC for Part: III item-16, with research question-1 groupings

Research Question-2

Is there a relationship between experiencing happiness and high school students' assessment of robotics and programming as interesting and cognitively developing?

The questionnaire item-2 in Part: II asked students for their assessment of robotics and programming as being interesting and helping to develop cognitive thinking. Therefore, this item was used to generate the groupings to compare as Part: III Happiness levels. The 'no robotics usage' cohort was group-1 (blue); those that answered 1, 2 or 3 on the Likert scale were placed in group-2 (red). Likert scale-4 and 5 answers were placed in group-3 (green) and level-6 and 7 responses were group-4 (purple). Plotting these groups against the Part: II, item-2 ICC shows the relative position against the model estimate curve, with those that strongly agreed with the statement (group-4) generally above the estimate, group-3 that agreed with the statement tracking the estimate and the other two groups below the curve (Fig. 11).

The table below shows the average happiness level logit value from Part: III for each of the four groupings for robotics programming being interesting and developing cognitive skills in Part: II (Table 3).

The data showed a general increase in overall happiness, with student agreement on robotics being interesting. The 'no usage group' has an average logit value of 0.433, the 'agreed group' having a logit value of 0.539 strongly agreed group scoring 0.636. Note that the 'disagree group-2' scored lower than the 'no usage group' at only 0.390.

Plotting these groups against individual items illustrates how each group compared to the model estimate curve for that model. Again using item-16 on the Part: III happiness survey for consistency, the strongly agree group (purple) and the agree group (green) plot above the model estimate and the no usage group (blue) generally

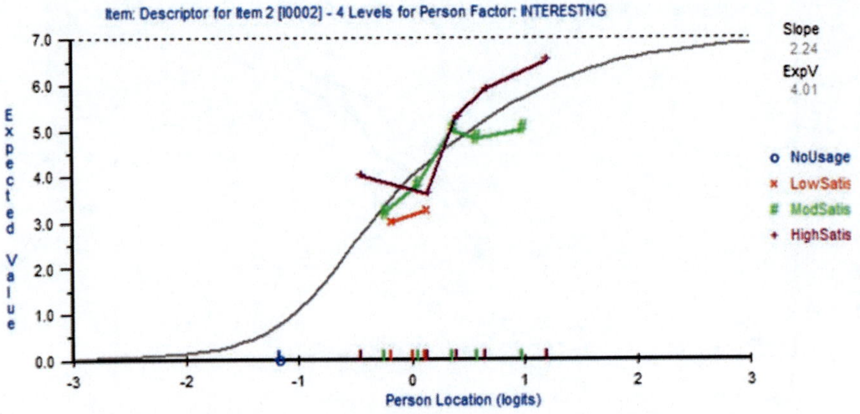

Fig. 11 ICC for Part: II item-2, with research question-2 groupings

Table 3 Part: III Happiness levels—interesting and cognitive

Part: III Happiness levels	
RQ 2: Interesting and cognitive	
Value	Logit average
No usage	0.433
Disagree	0.390
Agree	0.539
Strongly agree	0.636

plots below the estimate curve. The 'disagree group' (red) plots typically on the model estimate curve (Fig. 12).

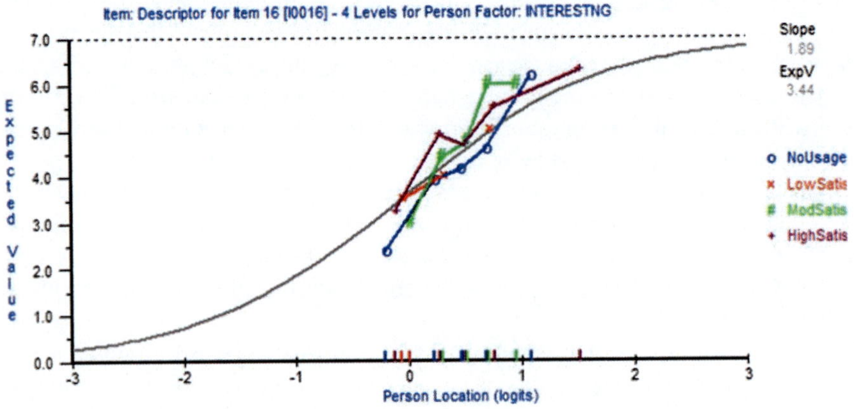

Fig. 12 ICC for Part: III item-16, with research question-2 groupings

Research Question-3

Is there a connection between experiencinghappiness and high school students' assessment of robotics classes as being useful in the context of creating knowledge in other subjects?

Part: II item-3 asked students for their assessment if robotics programming was useful for creating knowledge in their other subjects. Therefore, this item was used to generate the groupings to compare as Part: III Happiness levels. The 'no robotics usage' cohort was group-1 (blue); students that did not agree with this concept-1, 2 or 3 on the Likert scale were placed in group-2 (red). Likert scale-4 and 5 answers were placed in group-3 (green), and level-6 and 7 strongly agree responses were group-4 (purple). Plotting these groups against the Part: II, item-3 ICC illustrates the conformance of the strongly agreed group (purple) although a relatively consistent value above the estimated curve, the agreed group (green) irregular, averaging close to the estimate curve and the other two groups generally below the curve (Fig. 13).

The table below shows the average happiness level logit value from Part: III for each of the four groupings for robotics programming, assisting with building knowledge in other Part: II classes (Table 4).

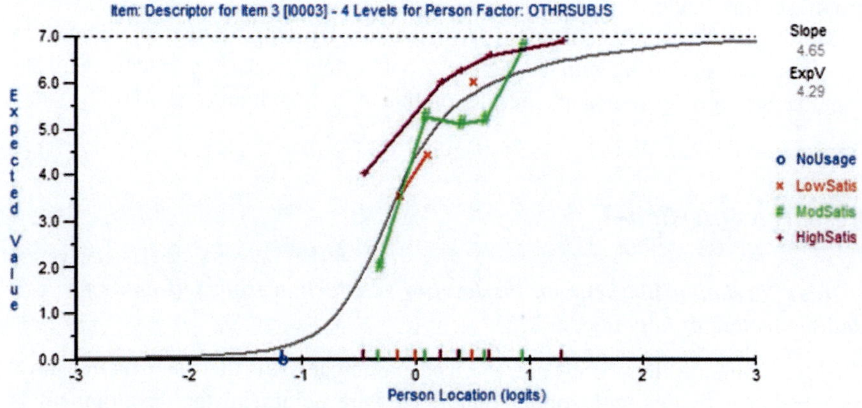

Fig. 13 ICC for Part: II item-3, with research question-3 groupings

Table 4 Part: III Happiness levels—other subjects

Part: III Happiness levels	
RQ 3: Other subjects	
Value	Logit average
No usage	0.433
Disagree	0.496
Agree	0.554
Strongly agree	0.656

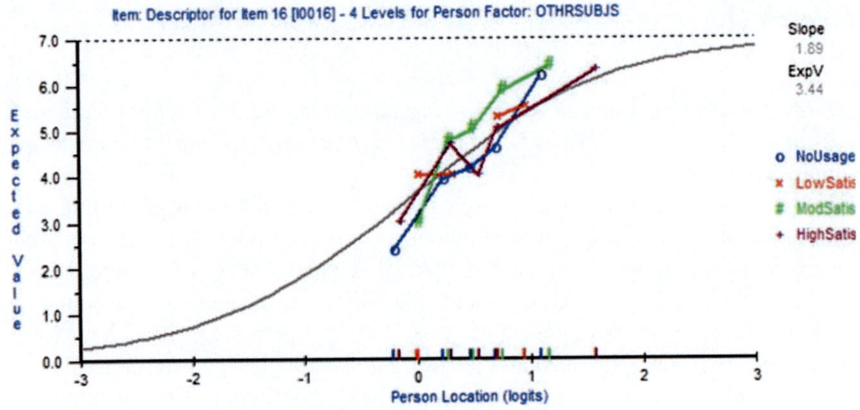

Fig. 14 ICC for Part: III item-16, with research question-3 groupings

The data shows a steady increase in the level of overall happiness with student agreement on robotics assisting with building knowledge in other subjects. The 'no usage group' has an average logit value of 0.433, the 'disagree group' a value of 0.496, the 'agreed group' having a logit value of 0.554 and the 'strongly agreed group' scoring 0.656.

Plotting these groups against item-16 on the Part: III Happiness survey indicated a partial reversal of scoring with the 'agreed group' (green) consistently scoring higher than the 'strongly agree group' (purple) on this individual item (Fig. 14).

Research Question-4

Is there a relationship between the level of satisfaction from self-made projects and the feeling of happiness?

The questionnaire item-5 in Part: II asked participants for their satisfaction level on whether activities with robotics and sets were helpful in the development of creativity and designing their own projects? Therefore, this question was used to group participants in answering the third research question. Four data groups were created: the group with no experience (group-1); the low satisfaction level (scores-1, 2 and 3 on the Likert scale, group-2); the moderate satisfaction group (scores-4 and 5 on the Likert scale, group-3); and the high satisfaction level group (scores-6 and 7 on the Likert scale, group-4). Group-3 has the opposite trend to the model estimate curve (Fig. 15).

The average logit score of each group from the filtered Part: III questionnaire was used to measure their happiness levels (Table 5).

The analysis again indicated a general increase in happiness levels with higher robotics satisfaction levels. The group with high satisfaction with their self-made

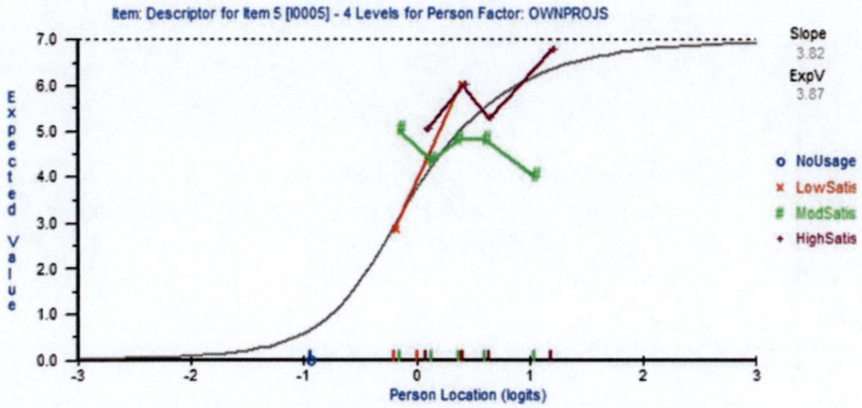

Fig. 15 ICC for Part: II item-5, with research question 4-groupings

Table 5 Part: III Happiness levels

Part: III Happiness levels	
RQ 4: Own robot project satisfaction	
Value	Logit average
No usage	0.390
Low satisfaction	0.462
Mod satisfaction	0.427
High satisfaction	0.805

projects had more than double the average logit value of the 'no usage' group, with the low and moderate satisfaction groups in between. Item-16 of the happiness survey indicates the 'no robotics group' (blue) generally had the lowest happiness scores along the model estimate curve. However, the other groups were not as clearly aligned on this particular item (Fig. 16).

Research Question-5

Is there a relationship between experiencing happiness and evaluating robotics and programming classes to develop teamwork by high school students?

Part: II item-6 asked students to assess robotics and programming to develop teamwork by high school students. This item was consequently used to generate the groupings to compare against Part: III Happiness levels. The 'no robotics usage' cohort was group-1 (blue); students that thought teamwork development was low (1, 2 or 3 on the Likert scale) were placed in group-2 (red). Likert scale-4 and 5 answers were placed in group-3 (green), and level-6 and 7 (strongly believing that teamwork

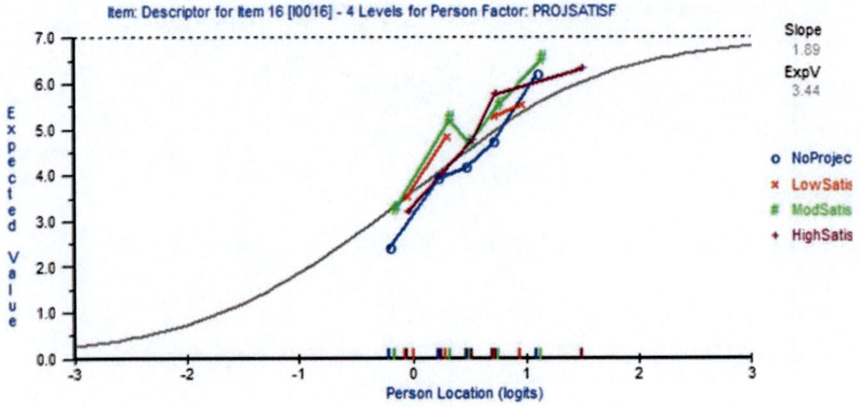

Fig. 16 ICC for Part: III item-16, with research question-4 groupings

was developed) were group-4 (purple). Plotting these groups against Part: II, item-6 ICC illustrated that the 'strongly agreed' group (purple) was generally highest above the estimated curve, but that the 'disagree group' (red) plotted higher than the agreed group (green). Both of these groups have an anomalously low observed value in the central zone of their distribution (Fig. 17).

The table below shows the average happiness level logit value from Part: III for each of the four groupings for robotics programming, assisting in developing high school students' teamwork from Part: II (Table 6).

The analysis had a general increase in happiness levels with a higher agreement in teamwork development. The group-4 strongly agreed that robotics helped high school students develop teamwork had the highest happiness logit average at 0.691, with the 'no usage group' at 0.440 and the 'disagree group' at 0.506. However,

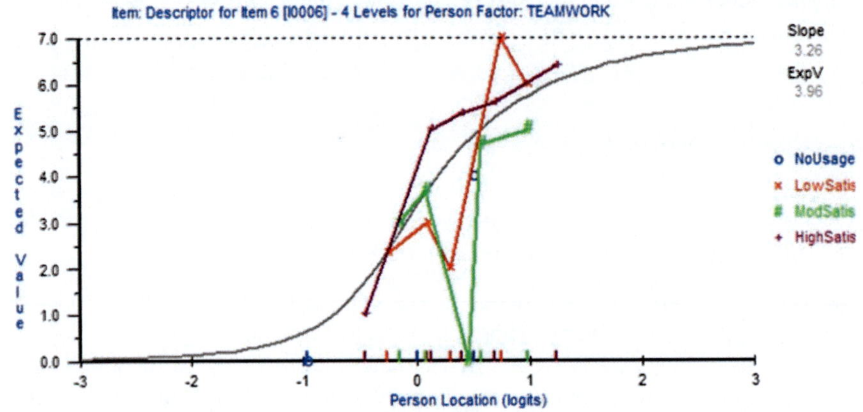

Fig. 17 ICC for Part: II item-6, with research question-5 groupings

Table 6 Part: III Happiness levels—promotes teamwork

Part: III Happiness levels	
RQ 5: Promotes teamwork	
Value	Logit average
No usage	0.440
Disagree	0.506
Agree	0.406
Strongly agree	0.691

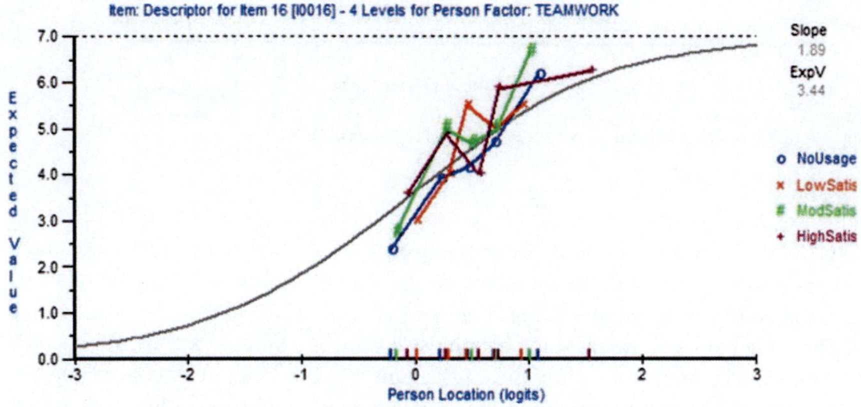

Fig. 18 Item characteristics curve for Part: III item-16, with research question 5 groupings

group-3 agreed that there was some benefit in developing teamwork with the lowest happiness logit score of only 0.406. This anomalous result was not reflected in Part: III item-16 ICC with group-3 (green) plotting above the estimate curve for this item (Fig. 18).

Research Question-6

Is there a relationship between experiencing happiness and evaluating robotics and programming classes to develop social competencies?

The questionnaire item-7 in Part: II asked participants if activities with robotics programming helped develop social competence? Therefore, this question was used to group participants in answering the sixth research question. Four data groups were created: the group with no robotics experience (group-1); students who did not think social competence was developed through robotics activities (scores-1, 2 and 3 on the Likert scale, group-2); students agreeing that some social competence benefit was achieved (scores-4 and 5 on the Likert scale, group-3) and students strongly

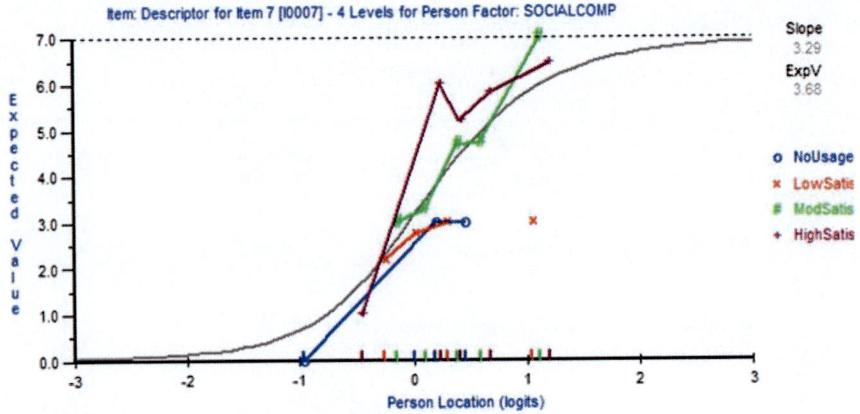

Fig. 19 ICC for Part: II Item-7, with research question-6 groupings

believing that robotics classes developed social competencies (scores-6 and 7 on the Likert scale, group-4). Plotting these groups on the Part: II item-7 ICC illustrates general conformance with the estimated model with the group hierarchy maintained across most of the distribution (Fig. 19).

The table below shows average happiness level (logit value) Part: III, for each of the four social competence groupings; maintains a general hierarchy concerning the 'no usage' and 'disagree' groups having average logit values in the low 0.4s, agree group at 0.534 and strongly agree group significantly higher at 0.703 (Table 7).

The ICC for Part: III item-16 illustrated some reversal of the groups with the 'disagree' group (red) plotting below the 'no usage' group (blue). The 'agree' (green) and 'strongly agree' groups (purple) were generally above the model estimate curve (Fig. 20).

Table 7 Part: III Happiness levels—social competence

Part: III Happiness levels	
RQ 6: Social competence	
Value	Logit average
No usage	0.442
Disagree	0.420
Agree	0.534
Strongly agree	0.703

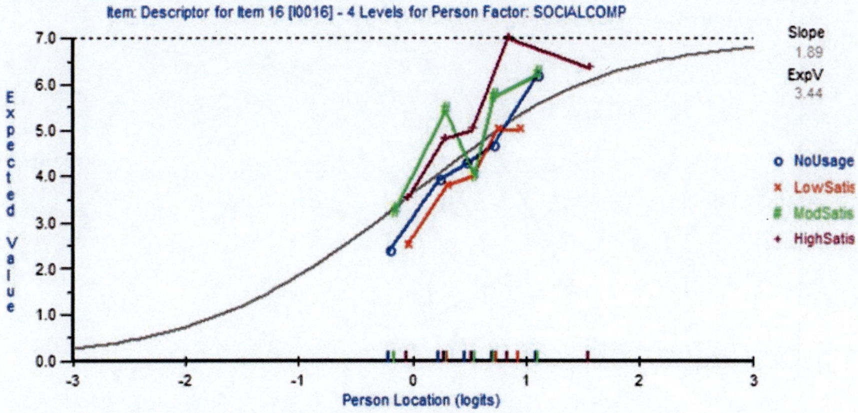

Fig. 20 ICC for Part: III item-16, with research question-6 groupings

Research Question-7

Is there a relationship between experiencing happiness and evaluating robotics and programming classes to develop communications' competencies?

Part: II item-8 asked participants if activities with robotics programming helped develop communication competence? This item was consequently used to generate the groupings to compare against Part: III Happiness levels. The 'no robotics experience' (group-1); students who did not think communication competence was developed through robotics activities (scores-1, 2 and 3, on the Likert scale, group-2); students that agreed that some communication competence benefit was achieved (scores-4 and 5 on the Likert scale, group-3) and students that strongly believed robotics classes developed communication competencies (scores-6 and 7 on the Likert scale, group-4). Plotting these groups on the Part: II item-8 ICC indicates a total reversal of the groupings, with the 'strongly believe' group (purple) the lowest on the plot and the 'no usage' (blue) and the 'disagree' group (red) above the model estimate (Fig. 21).

This anomalous result needs further investigation. The RUMM2030 ICC curve can be displayed with the observed values for the different class intervals. Figure 22 shows the ICC for item-7 and item-8. The observed values for item-7 plot along the Rasch model estimate curve with a fit residual of -1.002. However, for item-8, the observed value dots are scattered across the estimate curve, and the fit residual is poor at 4.508. There is an issue with either the question wording or the data collection in Part: II.

However, the table of 'communication competence groups' again illustrated the general increase in happiness levels. The average logit score increased from the 'disagree' group at 0.351, the 'agree' group at 0.588 and the 'strongly agree' group scoring 0.754. The 'no usage' group again scored higher than the 'disagree' group, as they did in research questions-2 and 6 (Table 8).

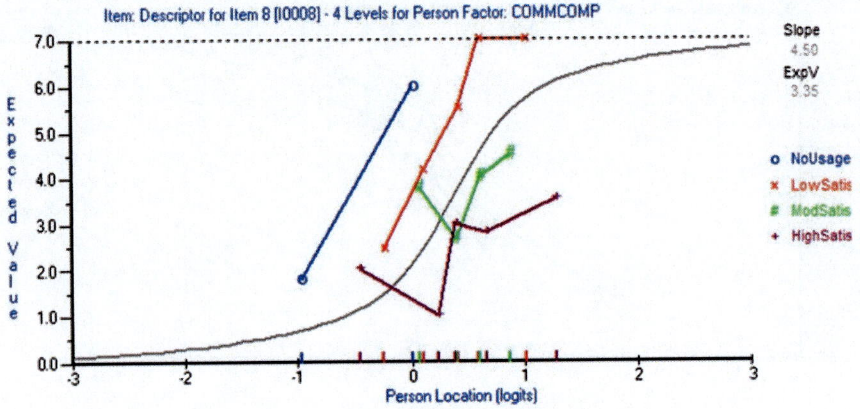

Fig. 21 ICC for Part: II item-8, with research question-7 groupings

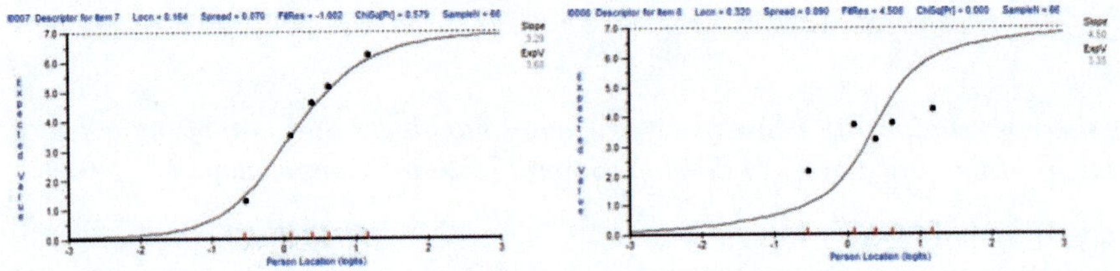

Fig. 22 ICCs with class observed values for Part: II item-7 and Item-8

Table 8 Part: III Happiness levels—Communication competence

Part: III Happiness levels	
RQ 7: Communication competence	
Value	Logit average
No usage	0.443
Disagree	0.351
Agree	0.588
Strongly agree	0.754

Plotting the 'communication competence' groupings on the usual Part: III item-16 ICC shows the same anomalous result. For this one item, the 'strongly agree' group plots the lowest against the model estimate, and the 'disagree' group was often above the model estimate (Fig. 23).

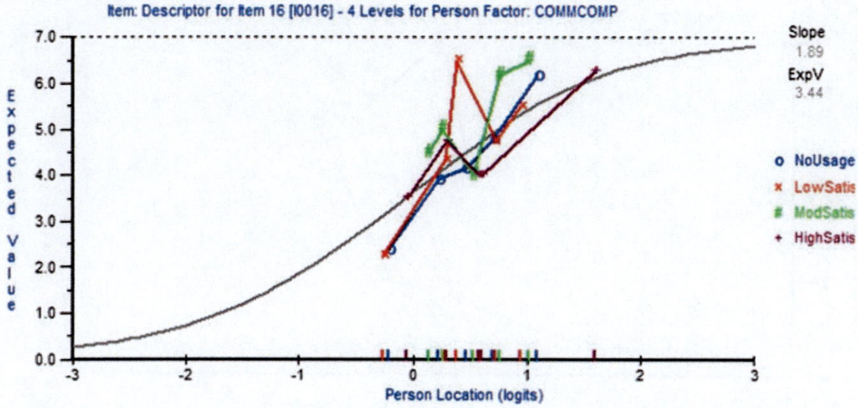

Fig. 23 ICC for Part: III item-16, with research question-7 groupings

Research Question-8

Is there a relationship between and the number of hours of classes with robotics and the feeling of happiness?

Items-14 to 18 in the Part: II questionnaire asked participants their satisfaction level on how often they had robotics programming or education kit experience in the classroom, ranging from once a month in item-14 (group-2) to daily in item-18 (group-6) (group-1 had no experience). To determine a participant's preference, the data analysts reviewed the students' answers to these five items. They grouped them based on which questionnaire item students selected to have the greatest satisfaction. If the students chose their highest satisfaction level for more than one item, they were grouped into the most frequent item with their highest score. The following ICC showed a distribution of the above six groups against the model estimate curve for item-17 (Fig. 24).

The table below shows the average logit score on the Part: III Happiness survey for each group (Table 9).

The analysis indicated a general increase in happiness levels with a preference for increased robotics class frequency. The two exceptions were that students preferring only monthly class exposure to robotics scored slightly lower than the group with no experience. The students who chose daily exposure to robotics scored 10% lower than students preferring robotics classes three times a week. The ICC for item-16 below illustrated the '3-times a week' group (pink), often scoring higher than the 'daily preference' group (brown) (Fig. 25).

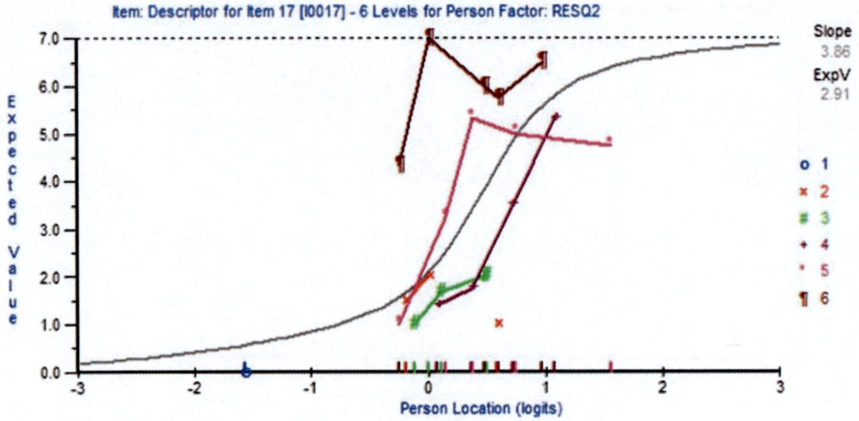

Fig. 24 ICC for Part: II item-17, with research question-8 groupings

Table 9 Part: III Happiness levels—class frequency

Part: III Happiness levels	
RQ 8: Class frequency	
Value	Logit average
No usage	0.378
Monthly	0.349
Weekly	0.541
Twice a week	0.536
3 Times a week	0.764
Daily	0.688

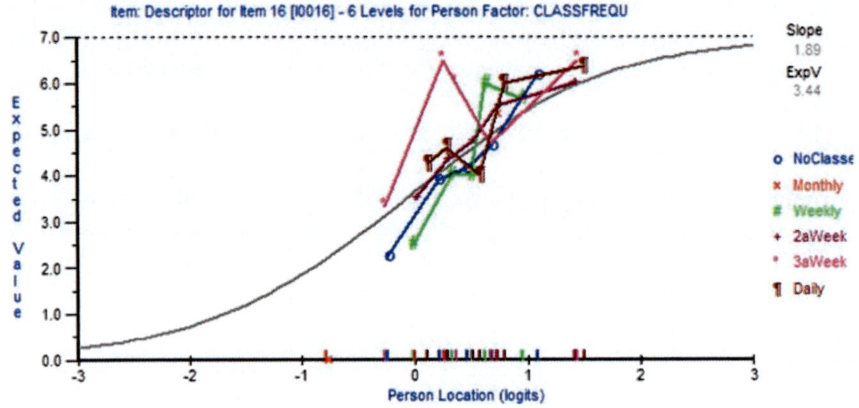

Fig. 25 ICC for Part: III item-16, with research question-8 groupings

Research Question-9

Is there a relationship between the prospect of a future job or career in robotics training and happiness?

The Part: II questionnaire had two items that referred to future job/career opportunities based on their robotics or educational sets. The data analysts selected item-20 to answer the research question-4 (*Please rate from 1 to 7 the prospects for further career and professional development depending on participation in classes with educational sets and robotics and your success in them*). Four data groups were created: the 'no experience' group; the 'low prospect level' (scores-1, 2 and 3 on the Likert scale, group-2); the moderate prospects group (scores-4 and 5 on the Likert scale, group-3) and the 'high prospects level' group (scores-6 and 7 on the Likert scale, group-4) (Fig. 26).

The average logit score of each group from the filtered Part: III questionnaire was used to measure their happiness levels (Table 10).

The analysis had similar end points to the previous research questions. The high job prospects group had more than double the average logit value of the 'no usage'

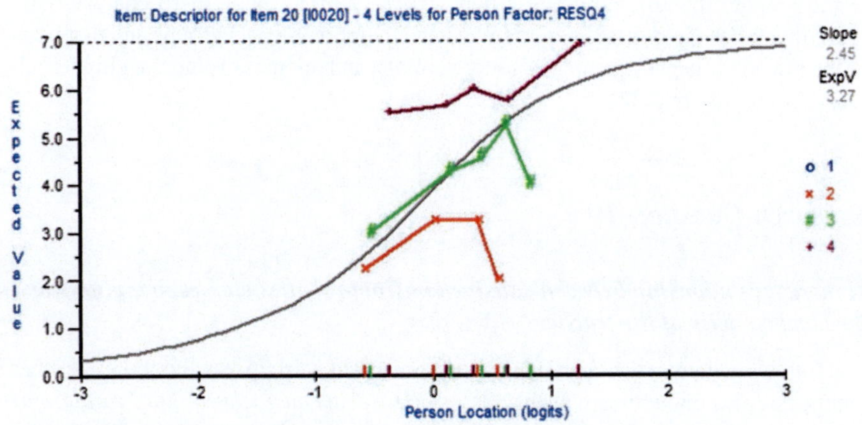

Fig. 26 ICC for Part: II item-20, with research question-9 groupings

Table 10 Part: III Happiness levels—future job with robots

Part: III Happiness levels	
RQ 9: Future job with robots	
Value	Logit average
No usage	0.381
Low prospects	0.610
Mod prospects	0.398
High prospects	0.764

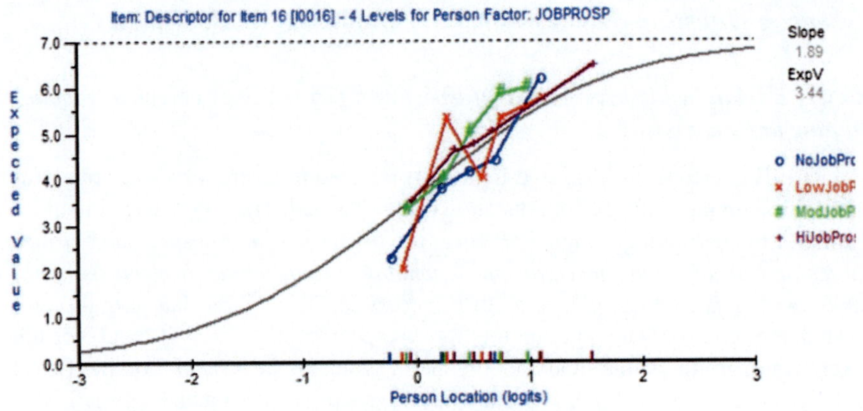

Fig. 27 ICC for Part: III item-16, with research question-9 groupings

group (happiness logit value of 0.764 compared to 0.381). However, the low job prospects group (0.610 average logit value, red) had a significantly higher happiness logit value than the moderate job prospects group (only 0.398, green), which was only slightly higher than the no usage group at 0.381, blue). The item-16 characteristic curve illustrates the group is generally below other job prospect groups and that parts of the low job prospect group (red) scored higher in happiness value than higher job prospect groups (Fig. 27).

Research Question-10

Is there a relationship between satisfaction from robotics classes, happiness and the competencies of the teacher?

The following item in the Part: II questionnaire asked participants about their teacher competency in conducting relevant classes (*Please rate on a scale of 1 to 7 rating your success on the teacher's interest and competence teaching classes with educational kits and robotics*). Four data groups were created: the 'no experience' group 1; the 'low competency level' (scores-1, 2 and 3 on the Likert scale, group-2); the 'moderate teacher competency' group (scores-4 and 5 on the Likert scale, group-3); and the 'high competency level' group (scores-6 and 7 on the Likert scale, group-4). The fit of each group against the model estimate curve was generally poor, but the group hierarchy was generally maintained (Fig. 28).

The average logit score of each group from the filtered Part: III questionnaire was used to measure their happiness levels (Table 11).

The analysis follows the same pattern as research questions 4, 5 and 9. The 'high teacher competence' group (purple) had more than double the average logit value of the 'no usage' group (blue) (happiness logit value of 0.815 compared to 0.381). The

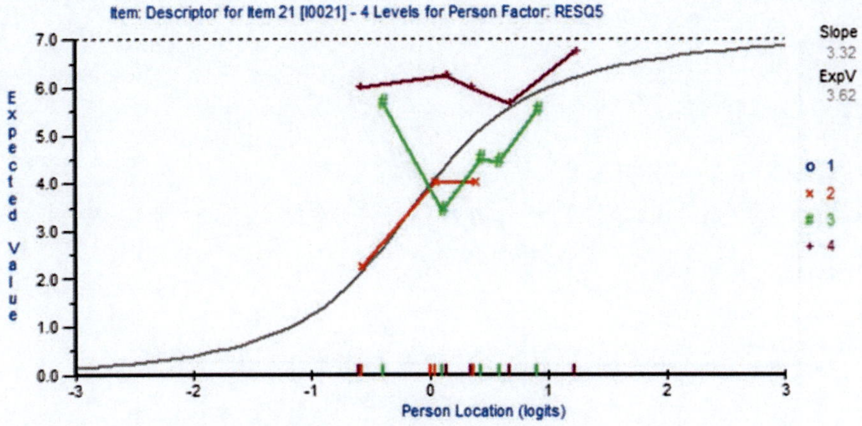

Fig. 28 ICC for Part: II item-21, with research question-10 groupings

Table 11 Part: III Happiness levels—teacher competency with robots

Part: III Happiness levels	
RQ 10: Teacher comp in robots	
Value	Logit average
No usage	0.381
Low competence	0.527
Mod competence	0.437
High competence	0.815

'low teacher competence' group (0.527 average logit value, red) had a marginally higher happiness logit value than the 'moderate job prospects' group (0.437, green) (Fig. 29).

Category Probability Curves

The RUMM2030 software can also produce probability curves for each item. Using item-16 from Part: III, the figure below shows the category probability curve (CPC), which illustrates the model estimate probability of a specific logit value response. This result is a cumulative probability so that each answer probability will add to 100% at any logit value. For instance, at a logit value of minus 2, the model predicts a 50% chance of the student answering 0, a 33% chance of answering 1, a 13% chance of answering 2, a 3% chance of answering 3 and a 1% chance of answering 4 (Fig. 30).

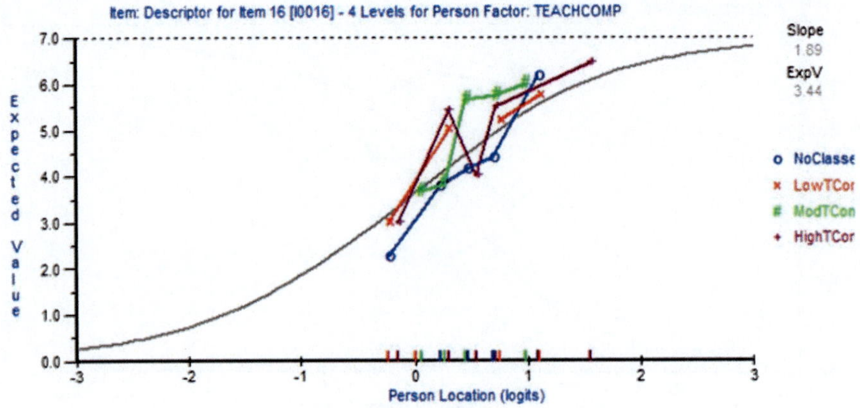

Fig. 29 ICC for Part: III item-16, with research question-10 groupings

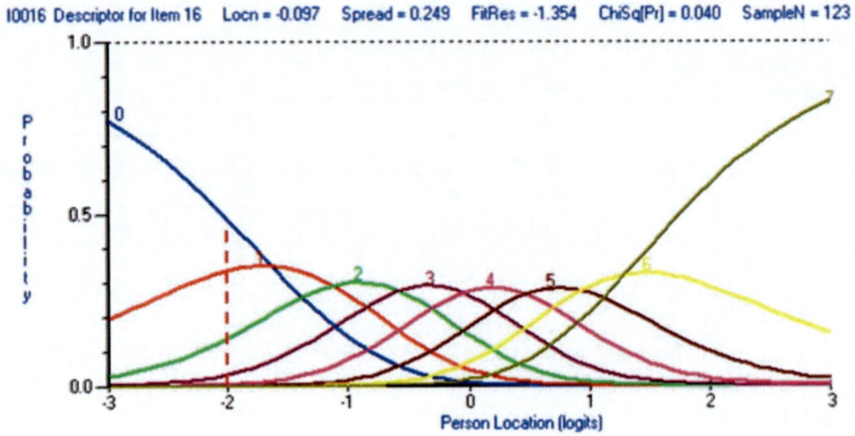

Fig. 30 CPC for Part: III item-16, showing cumulative probability estimate

Discussion

The RUMM2030 analysis of the Poland SDS data of 128 high school students participating in robotics programming classes delivers a robust outcome with logit value estimates for participants showing a systematic probability of happiness outcomes on individual questions. Figure 30 illustrates the increasing probability of higher Likert scale answers for happiness level as the test logit location of an individual increases.

This research involved a three-part study of collected data on the participant personal factors, their exposure and satisfaction to robotics programming in their classrooms and their general happiness levels. The literature showed terms happiness and well-being were understood differently across the globe. A question to ask might be whether a variable or continuum can be applied to happiness or well-being? The model implies a linear continuum and, as such, indicates that well-being is linear. Yet well-being is not always understood to be linear (Sumner, 2005). Initial data screening removed five participants whose answers were constant for all questions and therefore did not allow for any discrimination to analyse.

The personal factor data indicated that happiness increased from 0.320 logits for 15 year olds up to 0.622 logits for 18 years olds. The steady increase in happiness through ages 15 to 18 may reflect a building level of self confidence as they develop their knowledge and personal standing in the school. This is re-enforced by a similar increase from a Grade 1 average logit value of 0.530 to a top logit value of 0.751 in Grade 3. However, both factors show that happiness levels plummet for the senior graduating students, with the 19 year olds having an average logit value of only 0.380. The Year 5 students drop further to 0.143 logits, the lowest of all personal factors groups. The senior student happiness is likely to be negatively impacted by the pressures to obtain high grades for university entry and making important decisions about their future path in life, not to mention potential relationship issues. Village students are happier than city students (0.574 compared to 0.463, respectively) and secondary school students (0.587 logits) are happier than technical school students (0.488). This result may be related to secondary schools being more socially oriented while technical schools may focus more on technical training. There was no significant variation in happiness between males, females and LGBT participants.

The Part: II questionnaire was used as a screening tool to form sample groups for comparing on the Part: III Happiness test. The questionnaire asked students how their experience with robotics programming helped develop knowledge and various technical/social skills and how often they would like to have robotics classes. Ten research questions were developed from the 25-item test. Four groups were formed with the no robotics experience being one group and the other three groups derived from the Likert scale answers for nine of these questions. The remaining question related to how often the student had (or wanted) robotics classes. This question had six groups ranging from the no robotics experience to daily robotics classes. Plotting of these groups on the relevant Part: II item illustrated the general robustness of the RUMM model estimates. However, the plot for item-8 relating to the influence of robotics programming classes on developing communication competence indicated an issue with the item wording or corruption of the data collection.

The Part: III questionnaire was the focus of the research. This SDS questionnaire had 40-questions relating to general happiness levels of the students. These questions did not ask about happiness levels related to robotics programming. To link these two concepts, the groupings from Part: II were analysed on the RUMM analysis of the Part: III answers. The logit score of each individual on the Part: III test was captured and then grouped according to which group they were in for the ten research questions in Part: II. The statistical average of each group was then compared. There is a very strong correlation between happiness and satisfaction levels with all ten robotics programming question categories. Research Question-1 results are typical of the trend with students with no or limited robotics programming exposure having an average happiness level of 0.381 logits while students that are highly satisfied with their robotics programming experiences have a logit score of 0.762.

Conclusions

The earlier research of experts Donaldson et al., (2015) demonstrated that positive psychology is a growing and vibrant sub-area within the broader discipline of psychology, committed to using the same rigorous scientific methods as other sub-areas, in pursuit of understanding well-being, excellence and optimal human functioning (Donaldson et al., 2015) (p. 185). Other authors have stressed *"that workshops, using kits to build and program robots, are a modern form of interdisciplinary education of children and youth… as well as developing and training of key competencies"* of learners (Smyrnova-Trybulska et al., 2016). This topic is a modern trend and an important research direction. The current study indicated that robotics classes positively influenced learners' motivation to learn, promoted greater involvement in projects and increased their social and communication competencies with peers. Further comprehensive research on a larger, more diverse sample and with new variables is warranted.

This research moves beyond notions of homespun wisdom, weaving innovation, insight and experience together towards understanding deep learning attributes (https://deep-learning.global). The Rasch modelling shown here provides an exemplar to monitor how we shift from content-based learning (still predominant in Eastern studies) towards unfettered human-dimensional pedagogies, whereby teachers and students create learning experiences that build on students' innate curiosity and desire to solve complex issues and impact the world.

Moreover, this study has shown that robotics education encourages deep learning; armed with a well-founded interpretation of the Rasch estimate leads to improved equity issues not usually evident in classical test theory (CTT) commonly used for performance measurement. CTT estimates reliability by summing an individuals' total error score and their observed score to determine their total true score. Rasch model psychometric testing allows the researcher to explore beyond CTT, to adopt a more comprehensive unidimensional scale that empowers comprehension of human performance and test-item performance relative to each other. This methodology offers an effective tool for understanding individual differences in digital skill development.

Appendix—Research Survey

Research Questionnaire Addressed to Young Students of Secondary School/Technical School

Dear students! We kindly ask you and thank you in advance for completing the anonymous research questionnaire. It has 3 sections and is aimed at young people aged 16–20. The goal is to research the experience of happiness and satisfaction in robotics classes. Thank you for the filling and your time (about 10–15 min).

I. **SOCIOLOGICAL METRIX**

II. 1.1.How old are you?
    o 16, o 17, o 18, o 19, o other.

III. 1.2.What grade of high school do you study in?
    o 1, o 2, o 3, o 4, o 5, o Other.

IV. What type of secondary school do you study in?
    o secondary school, o technical school, o Other.

V. 1.4.Gender.
    o Woman, o Man, o Other.

VI. Where do you live?
    o Village, o City.

Part: II Survey

Questions related to your experience of participating in classes with robots and educational kits.

(1) Have you had any activities with robots and educational kits.
    o yes, o no.

(2) In the case of Yes, answer the next question: Rate on a scale from 1 to 7 the level of satisfaction (1 means complete dissatisfaction and 5 means the maximum level of satisfaction).

- Were the activities with robots and educational kits interesting and educational?
- Were the activities with robots and educational kits useful in terms of creating knowledge from other school subjects?
- Were the activities with robots and educational kits satisfactory?
- Were the activities with robots and educational kits helpful in developing creativity and designing your own projects?
- Were the activities with robots and educational kits helpful for the development of teamwork?
- Were the activities with robots and educational kits helpful in the development of social competences?

- Were the activities with robots and educational kits helpful in the development of communication skills?

(3) On a scale of 1 to 7, rate the level of satisfaction from the frequency of activities with robots and educational sets as additional activities:
o once a month, o once a week, o 2 times a week, o 3 times a week, o Every day.
(4) On a scale of 1 to 7, rate the level of satisfaction from the frequency of classes with robots and educational sets as compulsory classes in mathematics / computer science / technology lessons:
o once a month, o once a week, o 2 times a week, o 3 times a week, o Every day.
(5) What educational kits and robots were used during the classes? (Open).
(6) Rate on a scale from 1 to 7.

- the prospects of choosing your future profession related to the topic of classes with educational sets and robots and your success in them;
- prospects for further career and professional development depending on participation and success in learning kits and robots;
- conditioning your success on the interest and competence of the teacher who conducts classes with educational kits and robots
- for your future success from self-study with educational kits and robots at home (provided you have such kits)
- prospects for the development of future world technologies related to robotics and programming, artificial intelligence
- the teacher's personality and motivation to achieve better results in robotics classes
- the level of results obtained in robotics classes

NOTE: Parts of the above six questions were separated to form individual questions totalling 25 items for the data analysis.

Part III: The Scale of Experiencing Happiness (SDS)

This questionnaire contains a series of statements with which people usually describe themselves. They concern your experiences and feelings in various life situations. This scale provides information about the frequency of your happiness and the accompanying circumstances. Due to the fact that every person is different, there are no right and wrong answers, only answers that say something about you.

The results of this study are for scientific purposes only, so it is important that the answers you give are truthful.

We guarantee anonymity to all respondents.

Please read the following statements carefully and determine how each of them applies to you and your inner experiences. While completing the questionnaire, do not spend too much time on individual statements and do not go back to previous answers. To express your feelings, you have a 7-point scale at your disposal, described below.

Determine how often you are accompanied by the following sensations and sensations by marking one rating selected from this scale:

o 7 always, o 6 very often, o 5 often, o 4 sometimes, o 3 rarely, o 2 very rarely, o1 never.

Example:

I have a feeling of being in control and in control of my own life. 1 2 3 4 5 6 7.

you mark "7" if you always have this feeling.

In my life, I experience moments of maximum satisfaction. 1 2 3 4 5 6 7.

you mark "1" if you never experience that feeling.

A Research Questionnaire for Young People

Dear students! We kindly ask you and thank you in advance for completing the anonymous research questionnaire. It has 3 sections and is aimed at young people aged 16–19. The goal is to research the experience of happiness and satisfaction in robotics classes. Thank you for the filling and your time (about 10–15 min).

1. In my life, I experience moments of maximum satisfaction.
   Never, o 1, o 2, o 3, o 4, o 5, o 6, o 7, o Always.
2. I have a sense of control over my own attention.
   Never, o 1, o 2, o 3, o 4, o 5, o 6, o 7, o Always.
3. I consciously choose the goals which I am scrupulously pursuing.
   Never, o 1, o 2, o 3, o 4, o 5, o 6, o 7, o Always.
4. I perform various tasks only for the sheer satisfaction that comes from their implementation, and not to achieve some distant, external goals.
   Never, o 1, o 2, o 3, o 4, o 5, o 6, o 7, o Always
5. I can enjoy life despite adversities and unfavorable external conditions.
   Never, o 1, o 2, o 3, o 4, o 5, o 6, o 7, o Always.
6. Even the most ordinary events of everyday life can be pleasant for me.
   Never, o 1, o 2, o 3, o 4, o 5, o 6, o 7, o Always.
7. I can reward myself without reacting solely to the rewards and punishments distributed by society.
   Never, o 1, o 2, o 3, o 4, o 5, o 6, o 7, o Always.
8. I find satisfaction and purpose regardless of external conditions (social punishments or rewards).
   Never, o 1, o 2, o 3, o 4, o 5, o 6, o 7, o Always.
9. Sometimes I am "completely immersed" in the task at hand.
   Never, o 1, o 2, o 3, o 4, o 5, o 6, o 7, o Always.

10. While carrying out tasks, my emotions, intentions, goals and mind are in harmony with each other.
    Never, o 1, o 2, o 3, o 4, o 5, o 6, o 7, o Always.
11. I am completely focused on the activity I am doing, so that I lose track of time and there is no place in my consciousness for off-topic thoughts or feelings.
    Never, o 1, o 2, o 3, o 4, o 5, o 6, o 7, o Always.
12. I have moments when I feel active and strong.
    Never, o 1, o 2, o 3, o 4, o 5, o 6, o 7, o Always.
13. When faced with unpleasant sensations, I digest them for a long time and feel sorry for myself.
    Never, o 1, o 2, o 3, o 4, o 5, o 6, o 7, o Always.
14. In my life I try to clearly formulate my goals and strive to achieve them.
    Never, o 1, o 2, o 3, o 4, o 5, o 6, o 7, o Always.
15. I know exactly what actions to take to achieve the goals I am aiming at.
    Never, o 1, o 2, o 3, o 4, o 5, o 6, o 7, o Always.
16. I feel that I am making full use of my own skills when overcoming the difficulties I encounter.
    Never, o 1, o 2, o 3, o 4, o 5, o 6, o 7, o Always.
17. I take up challenges bordering on impossible.
    Never, o 1, o 2, o 3, o 4, o 5, o 6, o 7, o Always.
18. I feel that the demands of the external environment far exceed my skills and abilities.
    Never, o 1, o 2, o 3, o 4, o 5, o 6, o 7, o Always.
19. There are situations when I act "at full speed" being completely absorbed in a given activity.
    Never, o 1, o 2, o 3, o 4, o 5, o 6, o 7, o Always.
20. I strive to develop my own skills by increasing my expectations.
    Never, o 1, o 2, o 3, o 4, o 5, o 6, o 7, o Always.
21. I am willing to undertake activities that require significant skills and high concentration of attention.
    Never, o 1, o 2, o 3, o 4, o 5, o 6, o 7, o Always.
22. I spend my free time on active recreation (doing sports, hobbies, etc.).
    Never, o 1, o 2, o 3, o 4, o 5, o 6, o 7, o Always.
23. To come to relax I choose passive forms of rest that do not require a lot of "energy expenditure" (ie hanging around with friends, reading a not too difficult book, watching TV).
    Never, o 1, o 2, o 3, o 4, o 5, o 6, o 7, o Always.
24. I can get a lot of pleasure and satisfaction from my work.
    Never, o 1, o 2, o 3, o 4, o 5, o 6, o 7, o Always.
25. What I do is satisfying in itself, so I don't expect any external rewards for doing these activities.
    Never, o 1, o 2, o 3, o 4, o 5, o 6, o 7, o Always.
26. I can enjoy my activities.
    Never, o 1, o 2, o 3, o 4, o 5, o 6, o 7, o Always.

27. I have control over my own attention and over what I am going through.
    Never, o 1, o 2, o 3, o 4, o 5, o 6, o 7, o Always.
28. There are activities that I try to repeat, even though I have to put a lot of energy into doing them.
    Never, o 1, o 2, o 3, o 4, o 5, o 6, o 7, o Always.
29. I sometimes experience moments in my life that are incredibly "colorful and colorful".
    Never, o 1, o 2, o 3, o 4, o 5, o 6, o 7, o Always.
30. I am a person open to new, diverse experiences.
    Never, o 1, o 2, o 3, o 4, o 5, o 6, o 7, o Always.
31. I enjoy what I do, even if it is strenuous or difficult.
    Never, o 1, o 2, o 3, o 4, o 5, o 6, o 7, o Always.
32. I have a sense of being in control of my own life.
    Never, o 1, o 2, o 3, o 4, o 5, o 6, o 7, o Always.
33. When I feel that I am getting bored, I immediately find some work to counteract this feeling.
    Never, o 1, o 2, o 3, o 4, o 5, o 6, o 7, o Always.
34. I feel life is worth living.
    Never, o 1, o 2, o 3, o 4, o 5, o 6, o 7, o Always.
35. The matters that I deal with give me satisfaction, even when they are difficult and demanding.
    Never, o 1, o 2, o 3, o 4, o 5, o 6, o 7, o Always.
36. I can concentrate even for a long time on the performed activity.
    Never, o 1, o 2, o 3, o 4, o 5, o 6, o 7, o Always.
37. I am able to work hard.
    Never, o 1, o 2, o 3, o 4, o 5, o 6, o 7, o Always.
38. In my life I follow the principle that "it is never too late to learn".
    Never, o 1, o 2, o 3, o 4, o 5, o 6, o 7, o Always.
39. I try to introduce as many changes as possible into my life.
    Never, o 1, o 2, o 3, o 4, o 5, o 6, o 7, o Always.
40. I try to improve the ways of performing frequently repeated activities.
    Never, o 1, o 2, o 3, o 4, o 5, o 6, o 7, o Always.
    Thank you for answering the questions.

References

Andrich, D. (2011). Rating scales and rasch measurement. *Expert Review Pharmacoeconomics Outcomes Research, 11*(5), 571–585.

Crowe, S. (2021). 10 robotics startups to watch in 2020. *The Robot Report*.

Devlin, H. (2015). Rise of robots: How long do we have until they take our jobs? *The Guardian* https://www.theguardian.com/technology/2015/feb/04/rise-robots-artificial-intelligence-computing-jobs.

Donaldson, S. I., Dollwet, M., & Rao, M. A. (2015). Happiness, excellence, and optimal human functioning revisited: Examining the peer-reviewed literature linked to positive psychology. *The Journal of Positive Psychology Dedicated to Furthering Research and Promoting Good Practice, 10*(3), 185–195. https://doi.org/10.1080/17439760.2014.943801

Engelmann, J. B., & Pogosyan, M. (2013). Emotiona perception across cultures: The role of cognitive mechanisms. *Frontiers in Psychology.* https://doi.org/10.3389/fpsyg.2013.00118

Guglielmelli, E. (2014). Social needs and market demand for robots [From the Editor's Desk]. *IEEE Robotics & Automation Magazine, 21*, 4.

Hill, E. (2016). What is 'being in the zone'?—the fascinating psychology of super productivity, *HuffPost*, https//What Is 'Being in the Zone'? -- the Fascinating Psychology of Super Productivity | HuffPost.

Jiménez, F. R., Ruge, I. A., & Jiménez, A. F. (2020). Modeling and control of a two-wheeled auto-balancing robot: a didactic platform for control engineering education. In *18th LACCEI international multi-conference for engineering, education, and technology: engineering, integration, and alliances for a sustainable development hemispheric cooperation for competitiveness and prosperity on a knowledge-based economy*, July 27–31, Virtual Edition. https://doi.org/10.18687/LACCEI2020.1.1.556.

McKay, E. (2008). *The human-dimensions of human-computer interaction: Balancing the HCI equation.* In J. B. N. Balacheff et al. (Eds.), *The future of learning series,* Vol. 3. IOS Press, ISBN: 978-1-58603-869-4, ISSN:1572-4794 259.

Morze, N., Strutynska, O., & Umryk, M. (2018). Implementation of robotics as a modern trend in STEM education. *International Journal of Research in E-Learning, 4*(2), 11–32. https://doi.org/10.31261/IJREL.2018.4.2.02cv45t

Plichta, P. (2017). Socjalizacja i wychowanie dzieci i młodzieży z niepełnosprawnością intelektualną w erze cyfrowej. Wydawnictwo Adam Marszałek. *[Socialization and education of children and youth with intellectual disabilities in the digital age. Toruń: Adam Marszałek Publishing House.]*

Pogosyan, M. (2016). How does culture affect our happiness: Happiness and its many tastes. *Psychology Today*, How Does Culture Affect Our Happiness? | Psychology Today Australia.

Redd, D. & Klonoski, J. (2019). *How robots change the world: What automation really means for jobs and productivity, Oxford Economics, June.* How Robots Change the World—What automation really means for jobs, productivity and regions (oxfordeconomics.com).

Rosanda, V. & Starčič, A. I. (2019). A review of social robots in classrooms: Emerging educational technology and teacher education. *Education and Self Development, 14*(3), 93–106.

RUMM-2030. (2015). *RUMM2030 getting started manual.* RUMM_Laboratory_Pty_Ltd.

Smyrnova-Trybulska, E., Morze, N., Kommers, P., Zuziak, W., & Gladun, M. (2016). *Educational robots in primary school teachers' and students' opinion about STEM education for young learners.* Paper presented at the Proceedings of the International Conferences on Internet Technologies & Society 2016; Educational Technologies 2016 (ICEduTech 2016) and Sustainability, Technology and Education 2016 (STE 2016).

Sumner, L. W. (2005). *Happiness. Encyclopedia of Philosophy.* Macmillan Reference.

Yang, D., Oh, E.-S., & Wang, Y. (2020). Hybrid physical education teaching and curriculum design based on a voice interactive artificial intelligence educational robot. *Sustainability, 12*(19), 1–14.

Elspeth McKay Ph.D., https://www.cogniware.com.auafter 25 years at RMIT University, Melbourne, Australia, is moving back into her private consultancy, Cogniware. Her Ph.D. is in computer science and information systems, from Deakin University, Geelong, Australia. She also holds further qualifications in instructional design, computer education and business information systems. She is passionate about designing effective e-learning resources for industry training/reskilling programmes and the education sector. Her project interests involve psychometric testing on the effectiveness of online learning environments. Her work involves developing specialist e-learning tools that implement digital transformation precision. She has published extensively in the research fields of HCI and educational technology, with more than

150 peer-reviewed research papers and several educational technology books, and has been awarded numerous awards for funded projects.

ORCID profile https://orcid.org/0000-0001-7547-9616

Keven Asquith is CEO of 3D-GEO Pty Ltd https://www.3d-geo.com; a geoscience consultancy based in Melbourne, Australia. Keven has an Honours Degree in Geology (University of Western Ontario, Canada, and a Diploma in Project Management (University of New England, Australia). He has over 30 years of experience modelling the earth's sub-surface for the petroleum, mining, geothermal and Carbon Capture and Sequestration sectors. Keven has also spent several years in Government organisations developing IT project methodologies, tools, templates and designing/coordinating Board and Steering Committee reporting. Analysing complex data sets has always been a critical component of his 40 year career.

Eugenia Smyrnova-Trybulska Dr. hab., Professor at the University of Silesia in Katowice, Poland. She is Head of Department of Humanistic Education and Auxiliary Sciences of Pedagogy (2012–2019), Faculty of Arts and Sciences of Education in Cieszyn, University of Silesia in Katowice, Poland.

Eugenia is Coordinator of the Faculty Distance Learning Platform and of the Theoretical and Practical Aspects of Distance Learning Conference (http://www.dlcc.us.edu.pl), Editor-in-Chief of the International Journal of Research in E-learning IJREL (www.ijrel.us.edu.pl), E-learning Series indexed in WoS https://us.edu.pl/wydzial/wsne/nauka-i-badania/serie-wydawnicze/seria-e-learning/, a coordinator of the IRNet project (www.irnet.us.edu.pl) and of numerous national and international educational and research projects. She is the author of more than 260 scholarly papers and monographs in the field of e-learning methodology, ICT, multimedia in education, SMART university, STEM and STEAM education, teacher training in the area of ICT and e-learning, and is an author and methodological consultant of more than 160 distance learning courses.

ORCID profile https://orcid.org/0000-0003-1227-014X

Anna Porczyńska-Ciszewska Ph.D., Assistant Professor at the Institute of Pedagogy, University of Silesia in Katowice (Poland). She graduated from the Faculty of Pedagogy and Psychology with a major in psychology and obtained her master's degree in 2001. She obtained her doctoral degree in 2006 at the Faculty of Pedagogy and Psychology, University of Silesia in Katowice. She specialises in the psychology of health and illness and intellectual disability issues. Her research interests focus on problems connected with the functioning of people with intellectual disabilities and their families from a positive perspective, basing empirical studies on special needs pedagogy and positive psychology. She combines her research activity with working as a psychological practitioner, mostly providing assistance to people with psychological problems, intellectual disabilities, and their families. She also participates in international clinical trials. She is the Vice-President of the Committee on Psychological Sciences of the Polish Academy of Sciences.

ORCID profile https://orcid.org/0000-0002-5012-7590

Tomasz Kopczyński Ph.D., Assistant Professor at the University of Silesia in Katowice, Poland and Lodz University of Technology, Poland. Co-author of EduMata and EduMatrix teaching aids for teaching coding. Reviewer and specialist for the Polish National Center for Research and Development in the panel Digital Education. Author of the civic project for the city of Bielsko-Biala 'Multimedia Education Centers,' which resulted in the establishment of ICT/STEM education centers in 10 schools. Co-founder of the Coderdojo group for the local community aimed at promoting ICT/STEM skills among young people. Speaker of the National Academy of Programming. Technical editor, reviewer in TPF's scientific publication 'Chowanna' and 'Open Education Studies'. Manager of 'PFR Innovation Designers' of the Grow with Google program for the University of Silesia in Katowice. Manager of two local 2015/2016 editions of the international project 'Hour of code,' aiming to promote learning the basics of programming in children. Author of courses and papers for remote education in the subject 'Computational thinking' in partnership with the National Center for Research and Development.

ORCID profile https://orcid.org/0000-0002-8573-282X

Social Media Analytics, Learning Analytics and Healthcare Industry: Risky Drinking

Ashir Ahmed, Jenny Martin, Elspeth McKay, David Towl, and Zac Haussegger

Abstract Much is written about social media analytics, learning analytics and health care industry research projects. However, it is rare to find all three domains in the same study, as each discipline is usually kept separate from another. The concern with this isolation approach is the probability of missing essential synergies, leading to a combined effect more significant than the sum of their particular products. Social media analytics may uncover key social media search terms that reflect the language used by groups of people engaging in risky drinking behaviour. At the same time, the view of behaviour through a psychometric testing lens may reveal other things. This chapter describes both social media analytics and learning analytics from the perspective of the health care industry. The chapter explains the intersection of social media analytics and learning analytics; in the first instance, the Rasch measurement model is used to discover the probability of agreement relating to human behaviour issues; a case study amplifies health services provision outcomes. The results from the social media analytics revealed that groups' risky drinking behaviour data collected by Talkwalker showed linguistic variations with the double meanings of words used, which affected results. The learning analytics of alcohol consumption included a post-hoc comparison using a Bonferroni correction t-test which revealed the mean of never-married people was significantly higher than married people. The significance

A. Ahmed (✉)
School of Social Sciences, Media, Film and Education, Melbourne, Australia
e-mail: azahmed@swin.edu.au

J. Martin
School of Arts, Federation University Australia, Ballarat, Australia
e-mail: Jm.martin@federation.edu.au

E. McKay
Cogniware, Melbourne, Australia
e-mail: e.mckay@cogniware.com.au

D. Towl
Access Health and Community, Victoria, Australia
e-mail: david.towl@accesshc.org.au

Z. Haussegger
Swinburne University, Melbourne, Australia

E. McKay (ed.), *Manage Your Own Learning Analytics*, Smart Innovation, Systems and Technologies 261, https://doi.org/10.1007/978-3-030-86316-6_5

of these findings demonstrates that data analysis should be open to using a mixture of data analytics tools to reach a finer-grained interpretation to model other complex problems.

Historical Overview

Awareness for social media use has seen a constant increase since the advent of Web-2.0 when people found that collaboration in a shared digital environment improved their everyday lifestyles. There was a real fascination for people to reach out freely for information of any sort. Once the attractiveness of this emerging pastime spread through various local community platforms, a free-wheeling media frenzy ignited a digital media transformation. There are multiple ways people engage themselves in social media. The most common are Facebook, Flickr, YouTube, Twitter, Tik Tok, Instagram and Wikipedia. An ever-growing plethora of activities tempts people into digital connectivity as they share thoughts and wishes via virtual communities. Anything goes in this newfound information sharing ideology from learning a new craft, asking for advice on selling a house and seeking medical advice for ourselves and family members.

This potential for increased global distribution did not continue unrecognised for long by bright-eyed entrepreneurs. Revised business models soon emerged to innovate novel marketing opportunities. According to van Dijck (2013), a dual analytical prism has become apparent. On the one hand, a techno-cultural micro-system arose, whilst the other encompassed social media's socio-economic aspects. Whilst programmers realised a common thread between the two user connectivity platforms, there is a distinct difference between the two micro-systems. The underlying system services for designing the human–computer interaction (HCI) embodies the HCI's machine-dimension for the knowledge navigation's technical aspects. At the same time, the human-dimension (McKay, 2008; van Dijck, 2013) affords the social aspects.

Common aspects of sharing such knowledge online now include liking, following, trending and favouriting, all of which have a powerful effect on marketing recommendation platforms that bid for our social, political/ideological information control (van Dijck, 2013).

The Intersection of Social Media Analytics and Learning Analytics

The term data analytics (also known as data analysis) (Techopedia, 2021) refers to the science of analysing raw data for people to make conclusions about specific data, subjected to various examination techniques and mechanical processes that are often automated. Data analytics also refers to qualitative and quantitative research techniques used to improve productive-effort effectiveness in specific fields. Such examples include artificial and computational intelligence; and computer-science (for instance) in medical-biological and Web application firewall (WFA) signal sciences, social studies and social interactions, traffic-flows, computer and power networks, digital forensics and theoretical data analytics tool developments (Tsihrintzis et al., 2019).

Social media analytics and learning analytics have similarities in providing informed decisions on critical business operations and evaluating customer/learner performance outcomes. Yet, there are fundamental characteristics that set them apart, as both types of data analytics involve different expectations for their use. It is necessary to examine them separately to understand their conversion points.

Social media analytics primarily aids marketing decisions by gathering data from recommendation platforms to uncover buyer behaviour and social networking trends from online places like Facebook, Instagram and LinkedIn. Such analytics aim to track online conversations that inform community views, which may otherwise remain hidden. The overall expectation is to improve profit margins for the business sector.

Learning analytics also involves several evaluation techniques to aid decision-making concerning human performance behaviours, such as: creating reliable assessment reports for individual learners, to know how they rate compared with other learners; to highlight students who may need extra support; to assist teachers in planning their support interventions for individuals and groups of learners; and for professional development teams when considering new courseware design and development. The overall outcome in this educational sense is to improve the quality of the learning/instructional outcomes.

Finding the intersection is revealed by examining the underlying theoretical analysis models used to determine standard measurement techniques. Social media analysis takes place anywhere people gather together to discuss particular events and circumstances. The digital environment provides many tools that collect such unstructured data. This behaviour is typical, where people seek agreement/disagreement. In this sense, the community may create commonly known statistics models to represent their data. They pursue useful answers with means, standard deviation, t-tests and univariate analysis of variance (ANOVA).

Figure 1 depicts this process whereby people try to define the problem (Ganis & Kohirkar, 2015). Community members reach out through social media to find a pragmatic solution for dealing with the COVID-19 pandemic. They may look virtually towards others for suggestions on where to buy favourite brands that have depleted stocks in their local shopping centres. Social media in the pandemic crisis provides consumers with powerful brand awareness, and the marketing sector's business models rely on visibility to survive. However, when researchers adopt psychometric testing and follow a Rasch measurement model to conduct their data analysis, the focus is on psychological processes. Figure 1 depicts the differences in methodologies.

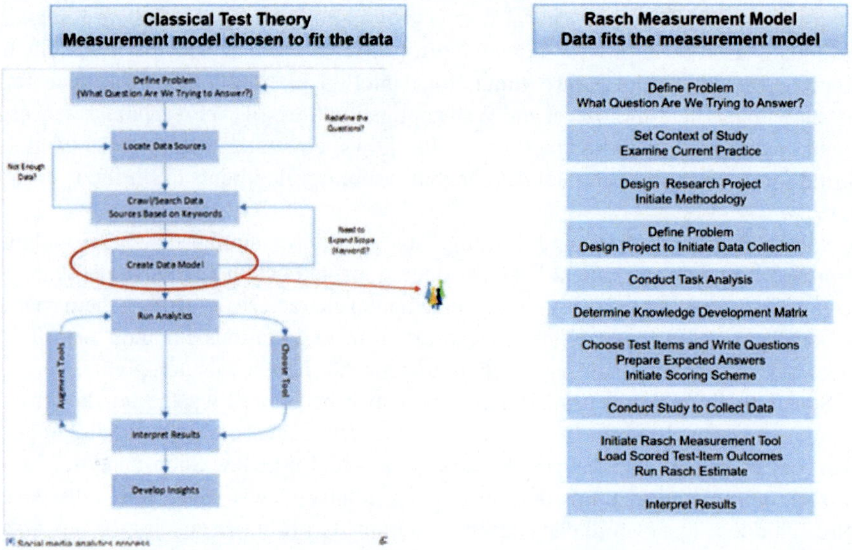

Fig. 1 Data analytics preparation models. (Adapted from Ganis & Kohirka, 2015)

Learning analytics typically involves psychometric (aptitude) testing to measure a person's suitability for a role, based on their intellectual capabilities and personality traits. Recruitment agencies in the business sector use psychometric assessments, believing that giving standardised tests will provide an objective assessment of who is the most suitable candidate. These tests, often called aptitude tests, are expected at job interviews; they include timed questions relating to maths, verbal comprehension and logic (diagrammatic) questions. The Rasch measurement theory was devised by analysing students' progress in learning (Rasch, 1960). George Rasch was a Danish mathematician, who developed models for analysing categorical data and investigating skill assessment and questionnaire/survey responses. Of course, there are always compromises between participant ability, attitudes or personality traits and test-item difficulty (Andrich & Marais, 2019). Therefore, Fig. 2 depicts the proverb that a picture speaks a thousand words as it communicates the initial research results using a graph effectively.

The starting point is to examine the results of a survey/questionnaire in terms of the individual participant or person-Likert agreement-scores relative to other participant's agreement-scores and the behaviour of other questionnaire-items, relative to each questionnaire item, on a common unidimensional scale.

Figure 2 shows the person-item threshold distribution involved 334-participants. The person threshold distribution shown as the pink-coloured histogram, with the test-items in the lower part of the graphic, demonstrated as the blue-coloured histogram. The summary statistics include the unidimensional Rasch logit-scale location (individuals and test-items), shown here between the person/item distributions and the lower placed fit-residual-statistics. As a concept, fit describes how well the

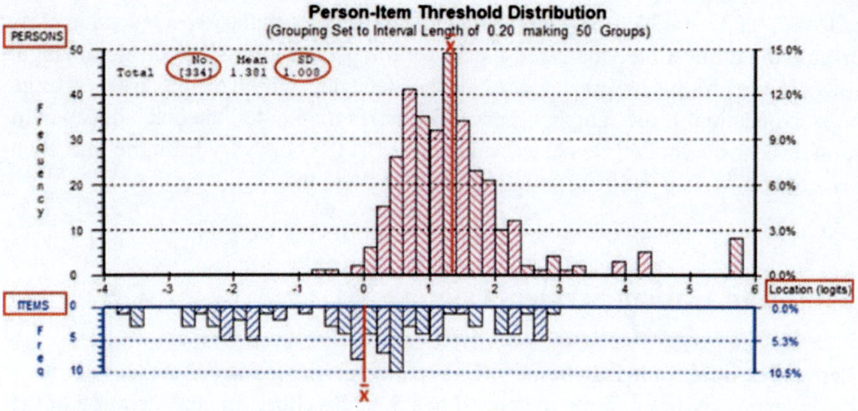

Fig. 2 A Picture speaks a thousand words

data conforms to the Rasch measurement theory. When the data does not fit, the model shows some divergence that the researcher must investigate. Therefore, the fit-residual shows the difference (large or small) as the standard deviation (Andirch & Marais, 2019: p 187).

Rasch-item characteristic curves identify class intervals that show an item's relative ability to discriminate amongst adjoining knowledge constructs (traits) along a linear scale (see Fig. 3). The learning analytics process shown here depicts a 5-level Likert scale corresponding to 1. never been used; 2. not effective; 3. effective to some extent; 4. effective; and 5. very effective. Rather than compared individual participant scores against the expected model curve, the Rasch model divides the sample into quartiles or the Likert scale quintiles, shown in Fig. 3. In doing so, it determines the mean logit score of each 20% of the participant sample. These means

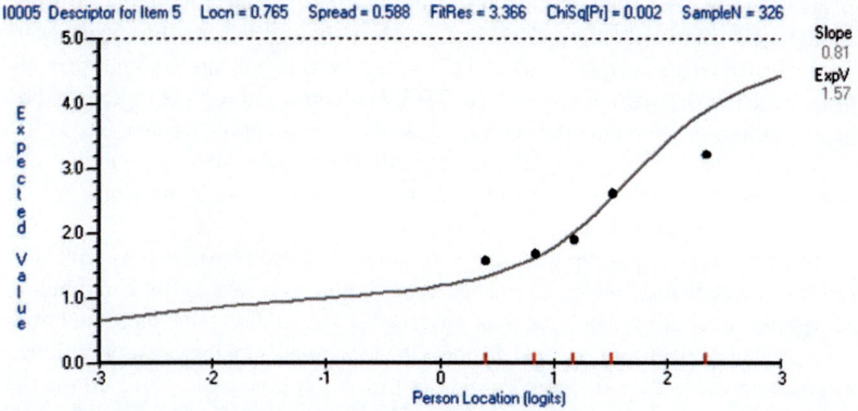

Fig. 3 Psychometric testing where data fit the model

are 0.402, 0.835, 1.175, 1.518 and 2.344. As you will see in the expected value plots for individual questions, these red marks on the person logit-scale do not change. However, the group's observed value (see the black dots within the plot) does change for each question. If the sample observed values fit the model, the black dots would sit on or near the model's expected value curve. If the dots vary from the line, then it is a poorer fit, and the fit-residual value will be larger.

Health and Human Services Context

There can be little doubt that the COVID-19 pandemic has given rise to many community awareness issues relating to general health and welfare. In consideration of the risks of the rapidly spreading virus, many governments imposed national lockdowns, where people had to remain at home, in an attempt to keep populations safe. It has illuminated many human behavioural issues including increased smoking and maintaining a balanced diet for keeping a healthy heart, bone density and immune system (University of Glasgow study finds that lockdown has had a 'major impact' on mental health—The Glasgow Guardian). In Australia, state governments issued fines for people ignoring the mandate to convey the importance of staying indoors.

In light of such community disasters, researchers in Greece conducted an online survey to detect alcohol consumption changes. They reported a four-day cross-sectional study based on convenience sampling because of the COVID-19 lockdown. The sample involved 705 adult health care professionals and was not representative of the general population. The mean age was 41.68, (standard deviation (SD) of 11.480); 75 per cent were female (528) and 25 per cent (177) were male. The mean male age of 43.10 (SD of 11.48) was a little higher than the mean female age of 41.21 (SD of 11.46); however, it was not significantly different 't 701 = 1.88, p > 0.001'. The ANOVA showed a significant difference between the pre-COVID-19 audit and marital status (Panagiotidis et al., 2020). Interestingly, the post-hoc comparisons using a Bonferroni correction t-test showed that the mean of never-married participants was 3.67 (SD of 1.57), which was significantly higher than the mean of the married participants as 3.10 (SD 1.41). This result was not repeated when the researchers explored marital status and alcohol consumption during the COVID-19 outbreak. Panagiotis et al. (2020) reported that participants who were living with their families reported lower alcohol consumption than people living alone '$t(702) = 2.91, p = 0.004, d = 0.21$'.

An observation regarding the metrics reported by these researchers is typical of traditional test theory, which estimates test-instrument reliability by summing an individuals' total error-score and their observed-score, to determine their total true-score. However, the Rasch model considers the individual's total score as a sufficient statistic to provide the individual's ability estimate. Psychometric testing allows the researcher to explore beyond Classical Test Theory, to adopt a more comprehensive unidimensional scale that empowers comprehension of human performance and test-item performance, relative to each other. This methodology offers an effective tool for understanding individual differences in digital skill development.

Case Study Healthcare Industry: Risky Drinking

Introduction of the Case Organisation

Access Health and Community (AccessHC) is a non-profit community health service in the inner east of Melbourne and played the role of the case organisation and funding body for this study. Aligning with its aims to be a leading primary health service provider in the area, and AH&C recognised that this project was an opportunity to transform the way it develops its interventions and deliver its services. As part of its innovation programme, the case organisation attempts to tap into the social media data to identify the social worlds of risky drinking culture in Victoria. The following sections outline the overview of the project, data collection method and the key findings of the study.

Overview of the Project

The drinking habits of people vary based on their social, cultural, geographical and professional status. In order to get a deeper and better understanding of social worlds of people engaged in risky drinking behaviour, this project examined social media data to understand the demographic, social and professional characteristics of the target populations in Victoria, Australia. For any attempt to assess or intervene in drinking practices, it is important to recognise that drinking alcohol is not a singular practice, but is rather intertwined within a range of other practices in what social practice theory describes as 'social worlds'. VicHealth and La Trobe University's Centre for Alcohol Policy Research (CAPR) (2019) define an alcohol culture as:

> *The way a group of people drinks, including the shared understanding of formal rules, social norms, practices, values and beliefs around what is and what is not socially acceptable when members of the group get together.* (p. 6)

The project included two phases. Phase 1 involved the sourcing of key search terms that potential risky drinkers were using when referring to alcohol on social media. This phase was divided into two parts. Search terms were generated from: (i) a scoping literature review and (ii) online surveys. Phase 2 of the project involved social media data collection using the search terms generated from Phase 1. The aim was to find risky drinkers in the wider population of Victoria through social media analysis. Phase 1 of the project answers the question:

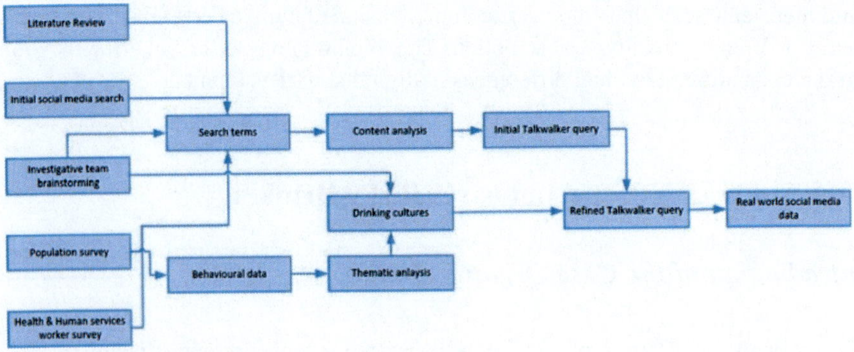

Fig. 4 Research design

What are the key social media search terms that reflect the language used by groups of people engaging in risky drinking behaviour? The process is outlined in Fig. 4.

The Scoping Literature Review

The role of scoping literature reviews is to inform policy decisionmakers when circumstances dictate a need for information in a timely manner (Langlois et al., 2017). Furthermore, scoping reviews have been shown to play a role when organisational and commissioning bodies see that a more succinct and timely process suits their own needs (Polisena et al., 2015). For these reasons, a scoping literature review of refereed articles was chosen as the most appropriate method. Initial searches were conducted by a senior librarian with most of the literature on this topic confined to the past ten years.

Literature Review Inclusion and Exclusion Criteria

Inclusion criteria were based around two core pillars: a focus on risky drinking behaviour and a focus on social media discourse. As such, terms used for the search included any of the phrases, 'risky drinking', 'harmful use of alcohol', 'alcohol misuse', 'drunkenness', 'intoxication' AND any of the phrases, 'social media', 'conversations', 'Facebook', 'Instagram'. To narrow the scope of our search, explicitly excluded items were those that centred around alcohol addiction and dependence and combined drug and alcohol use.

Literature Search and Article Review

Two searches were conducted in May 2020, using the EBSCOhost, Proquest central, Informit, Sage Journals and Web of Science databases. The initial search was limited to articles that were published in or after 2015. This search yielded 35 articles for review. Upon an analysis of the abstracts for inclusion, 18 of these were included for full-text review, and of this 18, four articles remained in the final dataset. A further search was conducted on the advice of an expert librarian to extend the search period to include articles published in or after 2010. This second search yielded a further 35 articles. Of these, 15 underwent further assessment via full-text review, and following this, two more articles were included in the final dataset leading to a total of six.

The main reason for exclusion at full-text review was under the pretence that studies were attempting to associate social media use with risky drinking behaviour rather than explore the interconnectedness as a phenomenon. Twelve of the 25 articles that were excluded were done so on these grounds. A further 2 articles were excluded in a similar way, but they offered to correlate user generated social media posts about alcohol with risky drinking behaviours. These articles demonstrated the two core pillars of our inclusion criteria, but upon further investigation, they revealed no information on the types of conversations or potential conversations that took place on social media. To the best of our knowledge, we were confident that we had located all of the relevant published literature on this topic through adopting this process. Table 1 provides details of the included articles and search terms used in the article inclusion process and the further information about those articles, including the search terms identified:

It is apparent that there is a very small amount of data available in the current literature that points to exact search terms applicable for social media trawling in regard to risky drinking behaviours. Final inclusion required a content analysis of the full-text articles to see if they contained relevant alcohol-related search terms. A short list of terms came up within several of the articles, with the term 'drunk' appearing in four of the six articles that provided search terms. The terms 'alcohol', 'cocktail(s)', 'drink(s)', 'vodka', appeared in three articles, whilst the terms 'binge drinking', 'drinking', 'get drunk', 'messy', 'rum', 'wasted' and 'wine' appeared twice.

Themes

In addition to the search terms identified, two main themes arose from the included articles; (a) positive and negative valence of social media posts and (b) adolescents, alcohol and social media.

Table 1 Articles included and search terms

Article citation	Method	Sample size	Country/Demographics	Search terms
Brown and Gregg (2012)	Author's own examples from their friend's social media pages	N/A	Australia/United Kingdom young adult women	Blossom hill, bombs, buca's, cocktails, crackin night, drink, drinkies, Friday feeling, girls gone wild, good times, jagerbuca's, jelly shots, last night, messy, mortal, never drinking again, out with the girls, ready for the weekend, shots, social drinks, sourz, suffering today, total carnage, vodka, weekend starts now, wine
Chainey and Stephens (2016)	Discourse analysis of New Zealand youth perception of risky drinking	$n = 20$ participants	New Zealand 16–18 year olds	getting wasted, wasted, hammered, drunk
Crocamo et al. (2020)	Three waves of tweets across different months of the year were analysed for alcohol-related hashtags after removal of 'source noise' to ensure the tweets came from real users tweeting about binge drinking	$n = 1{,}044{,}400$ tweets	United States General Population	#alcohol, #alcoholic, #alcoholics, #bingedrinking, #botellon, #cocktail, #cocktails, #drinking, #drinks, #drunk, #drunkasfuck, #drunkennights, #drunkies, #getdrunk, #hangover, #nomorealcohol, #pubcrawl, #pubcrawling, #rum, #sorehead, #toomuchalcohol, #vodka, #wasted

(continued)

Table 1 (continued)

Article citation	Method	Sample size	Country/Demographics	Search terms
Hebden et al. (2015)	Search terms are taken from focus group interviews discussing the connection between risky drinking and social media use	$n = 13$ participants, $n = 3$ group discussions	New Zealand university students	On the chop, drinking games, crazy antics, coma, drink, drunk, goon bag, messy night, classier night, four pack, alcohol, smashed, off my face, drinking, amped up, retarded, wine, binge drinking, casual beer
Meng (2017)	The study analysed a 1% random sample of publically available tweets in United states over a one year period	$n = 608{,}809$ alcohol-related tweets	United States general population	Beer, drunk, winery, beers, cocktail, cocktails, tequila, alcohol, champagne, margarita, vodka, martini, margaritas, rum, booze, get drunk
Merrill et al. (2020)	Search terms were decided by the research team to focus on the phrase 'blackout'	$n = 4736$ blackout-related tweets over three-day time period	Global general population	Black out, blackout, blacking out, blacked out, blacks out

Positive and Negative Valence of Social Media Posts

All six of the included articles either recognised the presence of sentiment within alcohol-related social media posts or set out to explore these explicitly. Meng et al. (2017) identified that 'social media platforms have a great potential to capture public sentiment' (p. 2), and that the number of posts that painted alcohol consumption in a positive light was greater than those portraying alcohol in a negative light. Measuring sentiment was done in a variety of ways. Meng et al. (2017) did so using an online text analyser that looked for specific strings within tweets, such as the use of 'smiley faces'. Brown and Gregg (2012) set out to legitimise alcohol-related social media that comes with drinking. Boyle et al. (2017) surmised that understanding the sentiment of these posts has its importance, as perceptions of peer drinking norms is the greatest predictor of drinking behaviours (p. 63). Their study took the concept of sentiment further by exploring the style of social media posts across different platforms. In their population of undergraduate students ($n = 209$), they found that posts that set out to glamorise drinking behaviour were commonly found on Instagram, whilst those depicting the consequences of intoxication to be seen on Snapchat.

Adolescents, Alcohol and Social Media

Half of the articles that were included focused their research questions on adolescents and undergraduate university students. Hebden et al. (2015) highlighted the role alcohol plays in the formation of friendships and peer acceptance, and the way rapid uptake of social networking sites amplifies perceptions and also misperceptions of drinking cultures.

Literature Review Concluding Comments

There is a very limited amount of evidence in the literature that links specific terms and phrases used on social media to those engaging in risky drinking behaviours. As a result, the list of terms generated is built largely on conjecture. Even for those articles that used a relatively rigorous evidence-based approach to the body of their research, the terms they generated each possessed elements of subjectivity. Along with a paucity of research available, amongst the current literature, there is a significant gap in knowledge as to what specific search terms are successful in finding specific groups of people engaging in risky drinking behaviours. When it comes to social media, most research that attempts to address risky drinking behaviours will focus on adolescents and university students. With growing numbers of people outside this age bracket adopting social media platforms, more research is needed to identify how other groups of people behave online in reference to their drinking habits.

Nonetheless the research terms compiled in this review have been valuable in informing our investigation into risky drinking behaviours in Victoria. The search terms identified in the scoping review have been used to inform the design of the two surveys of (1) general population and (2) health and human services workers in Phase 2. The aim was to identify further search terms for the Talkwaker analysis.

Population Surveys

Online surveys were distributed to two populations in the project: (1) the general population and (2) health and human service workers working with risky drinkers. The general population was chosen as it casts a wide net to discover individuals engaging in risky drinking. It is indeed possible that people engaging in behaviours that match the definition of risky drinking in this study might not consider themselves 'risky drinkers'. It is well-known that assigning labels to people in this way can lead to denial and, in these circumstances, declining participation. Online survey invitations were distributed to local community groups within Victoria such as sporting clubs, early childhood centres and schools.

The inclusion of the worker population aimed to provide another perspective on risky drinkers, with key questions being asked that allowed the investigators to formulate an idea of the types of drinking cultures practitioners thought were most at risk. Online survey invitations were distributed to health and human service workers via networks of alcohol and other drug services and community services including social workers, general practitioners and counsellors.

Inclusion criteria for the general population survey were that participants had to be: (1) over the age of 18 years, (2) drink alcohol or be affected by another person's alcohol consumption (without the requirement of being a risky drinker, as outlined above) and (3) use social media. Similarly, for the health and human service workers, to participate in the survey they needed to work with clients who (1) are over the age of 18, (2) engage in risky drinking behaviours or be affected by another person's alcohol consumption and (3) use social media.

Data Collection

Both surveys were distributed in August of 2020 via the survey distribution software, Qualtrics, for two weeks, followed by a re-distribution for another two weeks.

Online Surveys

The survey for the general population contained 23 questions on demographic information, social media and drinking behaviours and possible search terms. Many of the questions had pre-formed answers listed in drop-down boxes so that data collected would have an element of uniformity. The social media and drinking behaviours section contained closed-ended questions such as, 'how many people would you normally drink with?' The drop-down box contained responses such as 'one other' or 'a group of people' and then an option to select 'other' and detail a more specific response. Open-ended questions were also used, often added on the end of more targeted questions. For example, after asking if participants recognised or had used certain phrases that could be considered search terms for the project. Participants were then asked if there were any other terms and a text box was supplied. In this way, the survey adopted the use of both open- and closed-ended questions.

In addition to the key search term questions, behavioural questions were intended to further understand the drinking cultures of participants. In the section on the search terms, the list of pre-formed search terms was provided from the literature review with additional brainstorming by the investigators. Due to the Coronavirus (COVID-19) pandemic during the project, the investigators added a question at the end of both surveys enquiring about specific terms or phrases that refer to drinking specifically during quarantine and lockdown periods. The second survey, for workers in the health and human services sector, contained 12 questions. The first section contained closed-ended questions that detailed the occupation, location and workplace setting of the participant. In the second section, participants were asked to consider the people they work with that they were particularly worried about in terms of risky drinking. Open- and closed-ended questions were used here in a similar way to that of the population survey.

Data Analysis

The results from the targeted questions about possible search terms and hashtags from both surveys were compiled for use in the social media analytical tool, Talkwalker. The role that Talkwalker had in this project was to take the search terms provided by the surveys (and the literature review) and find social media users who are using those terms when engaging in interactions online. Talkwalker can return the additional information of the user's demographics, sentiment of the post and themes alongside each data point it returns, but it is governed by the privacy policy of the social media platform the data exists within. That means that on Facebook, for instance, posts that are only available on public pages will be accessed, and the demographic and personal information that goes along with those posts will not be retrieved unless that particular user has allowed that in their privacy settings. Channel-based monitoring works by monitoring publically available social media channels, such as a particular Twitter handle, a Facebook page or an Instagram hashtag. It then analyses all posts on that channel over a selected timeframe for the presence of the search term. This means that, on a known channel that represents a particular population within Victoria, by age group or region or occupation or interest or any combination of these, the investigators can find those who might be engaging in risky drinking behaviours.

Survey Findings

The surveys yielded two results as they pertain to the project: (1) the identification of risky drinking populations and (2) potential search terms that can be used in the Talkwalker analysis.

General Population Survey

Fourteen results were recorded on Qualtrics, between 16th August and 22nd September 2020. Seven of these were filled out completely, and the remaining seven

Table 2 Social world information from the general population survey

Occupation	Locality	Gender	Age	Alcohol intake	Drinking behaviour	Social media activity
Professional	Melbourne (inner)	Female	41–60	Risky	Social	Moderate
Professional	Melbourne (outer)	Male	41–60	Risky	Social/personal	Moderate
Clerical/admin worker	Melbourne (inner)	Female	25–40	Safe	Social	Moderate
Professional	Melbourne (inner)	Female	41–60	Safe	Personal	Low
Manager	Melbourne (inner)	Female	41–60	Safe	Social	Moderate
Clerical/admin worker	Melbourne (inner)	Female	60 +	Safe	Social	Moderate
Manager	Melbourne (inner)	Male	41–60	Risky	Family	Moderate

were left blank. Demographic, drinking and social media behaviours are summarised in Table 2 for each of the seven participants. Alcohol intake was deemed risky or safe in line with our definition of risky drinking from the NHMRC (2019) guidelines.

Worker Survey

The 'worker' survey yielded 10 results in Qualtrics between 16th of August and the 7th of September 2020. Of the 10 surveys, 4 were completed in full. Three additional surveys were completed up to the point of the participant detailing their demographic information and did not include any information about groups of interest. The final three surveys were returned fully blank. Risky drinking populations identified from the 4 completed worker surveys are detailed in Table 3.

Table 3 Social world information from the worker survey

Occupation	Locality	Gender	Age	Social media activity
Community/personal service worker	Melbourne (outer)	Female	41–60	No
Homemaker	Melbourne (inner)	Female	25–40	Yes
Unemployed	Melbourne (inner)	Male	41–60	Yes
Unemployed	Melbourne (outer)	Male	41–60	Yes

Table 4 Final list of search terms. Red Talkwalker topics excluded from topic query after they returned too many results

#afterworkdrinks	Alcohol	Hair of the dog	Tipsy
#cheers	Beer	Hammered	Virtual drink
#cheerstotheweekend	Binge drinking	Happy hour	Vodka
#cocktailhour	Blotto	Inebriated	Wasted
#drinkdrankdrunk	Cocktail	intoxicated	Wine
#drunk	Cocktails	It's 5 o'clock somewhere	Wine not
#liftyourspirit	Drink	Legless	Wine o'clock
#pinotgrigio	Drinking	Messy	Wine time
#pubcrawl	Drinks	Pissed	Winetime
#quarantini	Drunk	Rum	Zoom drink
#saturdaynight	Foursies	Shitfaced	
#whynot	Get drunk	Smashed	

Source Copied with permission from: Ahmed et al. (2021, p. 15)

Search Terms

A list of 41 search terms was compiled as a result of both surveys. In order to limit the size of the data obtained from Talkwalker, as the contract limited the maximum number of posts that could be stored, the search terms needed to be selected based on merit and quality of potential results. For this reason, the topic query contained search terms from the surveys that satisfied either of the following criteria:

1. The term was prescribed by the investigators and was selected by respondents on multiple occasions
2. The term was not on the prescribed list but added additionally by the respondents for the general population survey.

The topic query for Talkwalker also included terms that appeared on multiple occasions in the literature as outlined above. The final revised list of search terms is in Table 4.

Talkwalker Implementation

The Initial Topic Query

The original topic query included all of the items above, separated by the Boolean operator 'OR'. The terms in red, however, were so popular that on an initial search they returned posts that outnumbered our allocated limit, which was 30,000 posts over the 3-month period. It appeared that these very general terms that referred to alcohol (e.g. alcohol, rum, wine, beer, drink) as well as terms whereby their reference to alcohol was down the list in terms of their primary meaning (e.g. messy, smashed) were returning posts that seemed to be outside the scope of our project. For this reason, those terms were removed from the topic query. Additionally, terms from the surveys that referred to specific drinking occasions and behaviours were then selected for by the investigative team. These included #bottomlessbrunch, #speakeasy, #afterworkdrinks and #dranks. Such terms aligned much more closely with the project aim of discovering the social worlds of drinking in Victoria. Talkwalker specific filters directed the search to only take in results from Victoria, Australia, from Facebook, Twitter, or Reddit.

Data Cleaning

Once the dataset was obtained and downloaded, measures were taken to ensure that posts were relevant to the project aim. The importance of the cleaning process cannot be understated. The investigative team realised early on that the word 'pissed' yielded many results whereby the user was trying to express anger rather than referencing intoxication with alcohol. In order to assess the quality of the results from this query, two random samples of 200 posts were taken from the entire dataset that was generated by the initial topic query. These results were from the 15th October 2020 to 15th December 2020 (the entire length of the data scraping campaign). Each sample was assessed by a member of the investigative team, and their recommendations for the adjustment of the query were tested and then implemented across the entire dataset.

Recommendations were based on the particular post's relevance to the project aim. This was mainly achieved by assessing whether or not the post was referring to drinking when it used the search term. This process was repeated a second time on the resulting, now smaller, dataset, with the refinements implemented from the first iteration. In this instance, members of the investigative team double handled the same sample of 200. This second data cleaning enabled the team to reach a saturation point whereby no new information was being found towards the end of this sample. In this sample, greater than 80% of posts were deemed to be relating to drinking behaviours by each investigator. An outline of this process is detailed in Fig. 5.

Final Topic Query

In keeping with the recommendations and testing that occurred during that data cleaning phase, the final topic query utilised the Boolean operators AND, NOT and AND so as to specify the context for the words 'pissed', 'happy hour' and 'legless'. The words excluded were taken directly from examples from the sampling done in the data cleaning process.

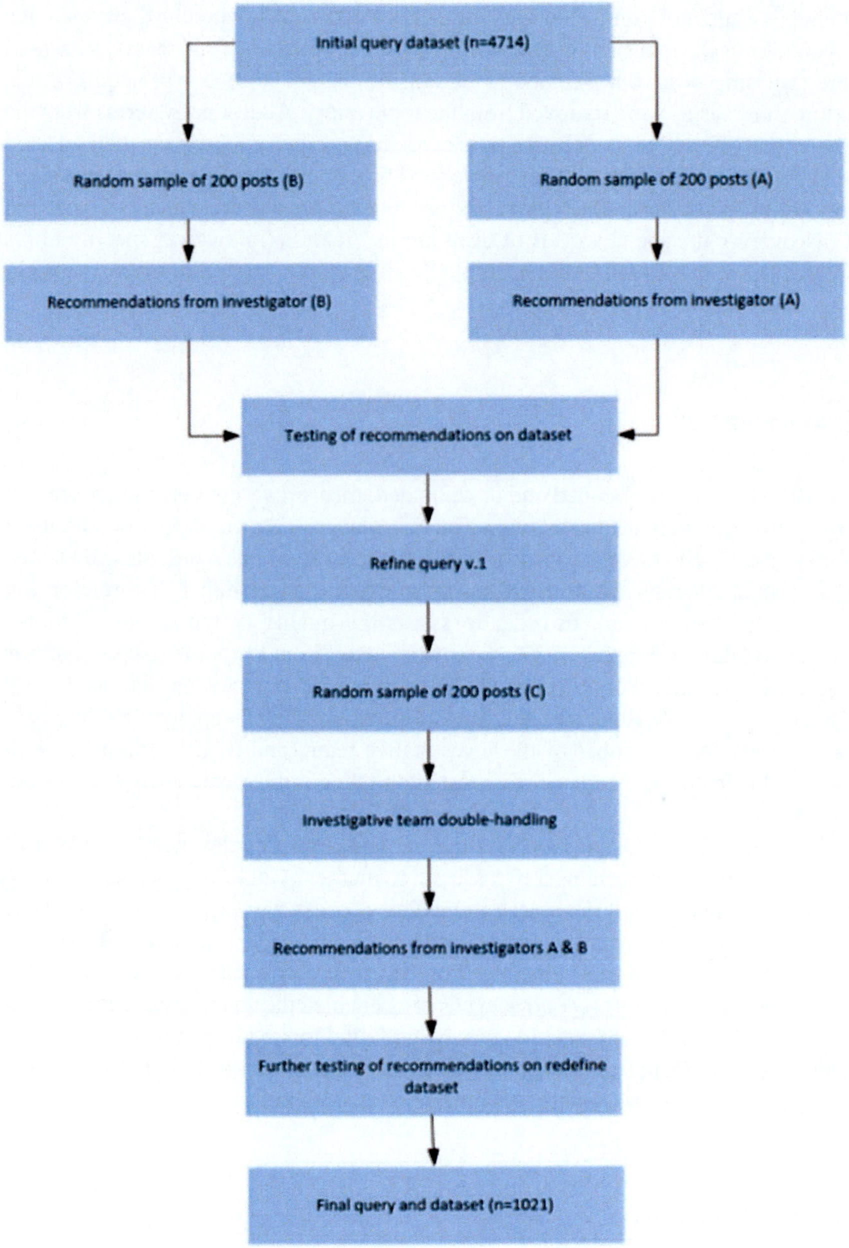

Fig. 5 Data cleaning process

Social Media Analytics

The complete dataset over the 3-month period included 4714 results. After the data cleaning process, the refined dataset included 1021 results. All of these results were from Twitter. Talkwalker has the ability to analyse a number of parameters pertaining to a user's profile when they find a post that matches keywords. Details such a location, age, occupation, gender and even interests can either be found explicitly on the user's public profile or be implied from their 'bio' or pages that they follow. Importantly, when analysing Talkwalker's summaries of the data, each parameter may have a different sample size. This is because on some user's profiles, the gender of the user may be available (either explicitly or implicitly), and on some others, it will not. At the same time, the occupation of the person might be available for the user whose gender could not be obtained but could not for the user whose gender could. Interpretation of the study findings must take this into account.

The social media data collected by Talkwalker suggest that parents from the age ($n = 578$) of 25–34 are involved more in risky drinking culture as compared to other age groups and other family status. In terms of gender ($n = 888$), males (55%) represent a greater proportion than females (45%). The most common occupation ($n = 405$) was authors or writers (18.3%), with the next one being photographers (6.9%). The top interests ($n = 1{,}129$) were found to be literature and books (8.5%) and food and drink (8.2%). Moreover, social media analytics suggests that top family status ($n = 899$) involved in risky drinking culture is parents (66.4%) and married as (29.3%). The full study findings are published elsewhere (Ahmed et al., 2021).

Conclusion, Limitations and Directions for Future Research

In addition to quantitative data analysis, a mini qualitative data analysis was performed during data cleaning process on the two samples of 200 posts.

Double Meanings in the English Language

In the first sample, when the dataset contained 4714 posts. About 138 posts had no drinking context. To fix this, the data cleaning process and manipulation of the topic query in Talkwalker ensued. This involved finding the words that had meant different things in different contexts, how they were used and with what other words they were used when not in a drinking context. Of these, 138,128 were due to the word 'pissed'. On only 5 occasions within, this sample did the word 'pissed' refer to drinking or being 'drunk'. For projects like this, in which the meaning of the language is so central to its methodology, the English language poses this problem. Further data cleaning in this way led to a final dataset of 1021 posts. The inherent double meaning of words has essentially taken up s posts available in the data limit of the project and should be accounted for and expected in similar projects in future.

Tweeting Style

Once the word 'pissed' (as well as a number of others) were accounted for and the query was adjusted, the second sample of 200 posts were taken from what turned out to be the final dataset. It remained the final dataset as 167 of the 200 posts were referring to words within the topic query in a drinking context. In this sample, posts were coded to assess the style of tweet that was occurring. The most common was a tweet that contained a reference to drinking in part of a twitter conversation between friends or online acquaintances, as either a reply or an initiation of a discussion. This occurred 49 times in the sample. The second most common tweet was one in which an image was attached. This occurred 38 times.

What were more concerning from this sample, in the context of the project—that is, one with a focus on 'risky drinking', were the posts referring to binge drinking ($n = 10$), blacking out ($n = 1$), day drinking ($n = 3$), having sex ($n = 1$) and drinking alone ($n = 3$). In contrast to some of the twitter 'conversations' as described above, some of these posts involved asking any of their followers to meet to have a drink with them somewhere without referring to anyone in particular.

There were also instances of posts that conveyed what the research team would deem as more responsible drinking. These included posts in which alcohol was consumed with a meal ($n = 6$) and alcohol involved in a moment of celebration ($n = 7$). In this sample, there were also posts that referred specifically to drinking during the COVID-19 pandemic ($n = 14$).

Sentiment

Talkwalker has the capability of analysing sentiment of the posts it obtains. It does this by assigning either a +5, 0 or −5 value to each post, correlating with a positive, neutral or negative sentiment, respectively. To assess sentiment, a random sample of 40 posts was taken for independent analysis of sentiment by two members of the research team. In 20 posts, Talkwalker had assessed the sentiment in the same way as at least one of the researchers. In fact, the research team members differed on their assessment of sentiment on 24 occasions.

Project Recommendations

The utilisation of Talkwalker to find people engaging in risky drinking behaviours is promising. In saying this, there is limited confidence in providing the social world of 25–34-year-old males, who are authors or writers, are parents, and enjoy literature and books, as the target of Victoria's next alcohol intervention strategy. The project does indicate, however, that in a world where social media is at the front of corporate consumer marketing strategy, it too can be at the front of public health intervention and strategy. The project has demonstrated that discussions around drinking do take place online, and there is capacity within software like Talkwalker to find them.

For a larger scale search within Victoria using this method, it would benefit from firstly adopting the technical learnings as demonstrated in this chapter. Secondly, it is important to be clear about the social worlds it is targeting and aim to utilise a more effective strategy for monitoring Facebook and other sites deemed appropriate. Thirdly, a manual qualitative data analysis on the raw data for a more informed search would reveal a more in depth reading of the drinking cultures and risky drinking behaviours in Victoria. What is promising, though, as datasets get larger, is that the sentiment algorithm within Talkwalker has the propensity to be fine-tuned to reveal risky drinking behaviours itself. Therefore, lastly, an adoption of machine learning in this way is recommended for future projects of this nature.

Summary

The primary purpose of the chapter was to highlight the advantages of using multiple data analytics techniques. Following this path affords a rich synthesis of usually disparate fields. Whilst there are many commonalities between social media analytics, it is the need for careful planning of the knowledge captured and the development of testing instruments for understanding the complexity of the data analysis (learning analytics).

The final case study outcomes revealed three interesting situations. These categories amounted to Talkwalker linguistic issues when people interpreted words differently. In so far as Tweeting style, the concerning example arose with attitudes on risky drinking, where posts were referring to binge drinking ($n = 10$), blacking out ($n = 1$), day drinking ($n = 3$), having sex ($n = 1$) and drinking alone ($n = 3$). At the same time, responsible behaviours towards responsible drinking included: consuming a drink with a meal ($n = 6$), involving alcohol in celebratory circumstances ($n = 7$) and posts from the same sample group regarding drinking during the COVID-19 pandemic (n = 14). For dealing with the sentiment category, Talkwalker uses a +5, 0 or −5 to correlate with a positive, neutral or negative sentiment, respectively; 40 posts were taken for independent analysis by two researchers, twenty posts aligned with one of the researchers. There were 24 occasions where research team members differed on their assessment of sentiment.

The overall contention of the case study was to recognise that drinking alcohol was not a singular practice; instead, it was intertwined within a range of other social practices described as social words. In Phase-1, the aim was to find the key social media terms that reflect the language they were using. This was done via a scoping literature review and the distribution of two surveys (1. general population and 2. health and human services). Phase 2 involved the utilisation of these search terms to find people engaging in risky drinking in the broader population.

Data analytics ventures across statistical pathways; on the one hand, the classical approach for finding a suitable model to undertake the interpretation restricts the researchers to follow rigid rules. On the other hand, psychometric testing provides a proactive approach where the researchers focus on a complex statistical environment that supports the measurement of diverse human traits. Health and human services lie at the intersection of social media analytics and learning analytics. This is where the Rasch measurement model affords both statistical rigour and the researcher's desire to uncover hidden reasons for interpretations of human behaviour through validated instrumentation. The individual's probability for expected outcome/agreement levels is measured with certainty and may be predictable.

References

Andrich, D., & Marais, I. (2019). *A Course in rasch measurement theory: Measuring in the educational, social and health sciences*. Springer Texts in Education.

Brown, R., & Gregg, M. (2012). The pedagogy of regret: Facebook, binge drinking and young women. Continuum. *Journal of Media & Cultural Studies, 26*(3), 357–369. https://doi.org/10.1080/10304312.2012.665834.

Chainey, T. A., & Stephens, C. (2016). 'Let's get wasted': A discourse analysis of teenagers' talk about binge drinking. *Journal of Health Psychology, 21*(5), 628.

Crocamo, C., Viviani, M., Bartoli, F., Carra, G., & Pasi, G. (2020). Detecting binge drinking and alcohol-related risky behaviours from Twitter's users: An exploratory content- and topology-based analysis. *International Journal of Environmental Research and Public Health, 17*(5).https://doi.org/10.3390/ijerph17051510.

Ganis, M., & Kohirkar, A. (2015). *NYSocial media analytics: Techniques and insights for extracting business value out of social media*. IBM Press.

Hebden, R., Lyons, A. C., Goodwin, I., & McCreanor, T. (2015). When you add alcohol, it gets that much better': University students, alcohol consumption, and online drinking cultures. *Journal of Drug Issues, 2*(214). 1177/0022042615575375.

Langlois, E. V., Starus, S. E., Mijumbi-Deve, R., Lewin, S., & Tricco, A. C. (2017). The need for rapid reviews to strengthen to inform health policy and systems. In E. V. L., A. C. Tricco, & S. E. Straus (Eds.), *Rapid reviews to strengthen health policy and systems: A practical guide*. World Health Organization.

McKay, E. (2008). *The human-dimensions of human-computer interaction: Balancing the HCI equation* (1 ed. Vol. 3). IOS Press.

Meng, H. -W., Kath, S., Li, D., & Nguyen, Q. C. (2017). National substance use patterns on Twitter. *PLoS ONE1, 12*(11), 1–15. https://doi.org/10.1371/journal.pone.018769.

Merrill, J. E., Ward, R. M., & Riordan, B. C. (2020). Posting post-blackout: A qualitative examination of the positive and negative valence of tweets oosted after 'blackout' drinking. *Journal of Health Communication*. https://doi.org/10.1080/10810730.2020.1719242

Panagiotidis, P., Rantis, K., Holeva, V., Parlapani, E., & Diakogiannis, I. (2020). Changes in alcohol use habits in the general population, during the COVID-19 lockdown in Greece. *Alcohol and Alcoholism, 55*(6), 702–704. https://doi.org/10.1093/alcalc/agaa092

Polisena, J., Garritty, C., Kamel, C., Stevens, A., & Abou-Setta, A. M. (2015). Rapid review programs to support health care and policy decision making: A descriptive analysis of processes and methods. *Systematic Reviews, 4*(1). https://doi.org/10.1186/s13643-015-0022-6.

Rasch, G. (1960). *Probabilistic models for some intelligence and attainment tests.* Nielsen & Lydiche.

Techopedia. (2021). Data analytics. https://www.techopedia.com/definition/26418/data-analytics.

Tsihrintzis, G. A., Sotiropoulos, D. N., & Jain, L. C. (2019). *Machine learning paradigms advances in data analytics.* Springer International.

van Dijck, J. (2013). *The culture of connectivity: A critical history of social media.* Oxford University Press.

Ashir Ahmed is a senior lecturer of media data analytics at Swinburne University of Technology, Australia. He earned his Ph.D. in Information Systems from Monash University. He has been involved in research projects related to data visualization and business analytics for non-profit organizations. His other research projects include the use of social media for suicide prevention, evaluating the role of community consultation for anti-radicalisation through e-education, and developing the framework for using web 3.0 in disaster management.

Jenny Martin is Professor of Social Work and Human Services at Federation University Australia. Jenny is well experienced in the field of mental health as a social work researcher, educator and practitioner. Current research is focused on inter-disciplinary collaboration across health disciplines to improve clinical communication skills and client outcomes. This includes effective use of technology to enhance inter-disciplinary collaboration. Her research into workforce development in social work, mental health and aged care focuses on issues of access and equity highlighting the importance of translational research to inform practice outcomes and the importance of communicating this in reputable peer reviewed publications as well as to policy makers and in practice forums.

ORCID profile https://orcid.org/0000-0001-7915-3138

Elspeth McKay Ph.D., https://www.cogniware.com.auafter 25 years at RMIT University, Melbourne, Australia, is moving back into her private consultancy, Cogniware. Her Ph.D. is in computer science and information systems, from Deakin University, Geelong, Australia. She also holds further qualifications in instructional design, computer education and business information systems. She is passionate about designing effective e-learning resources for industry training/reskilling programmes and the education sector. Her project interests involve psychometric testing on the effectiveness of online learning environments. Her work involves developing specialist e-learning tools that implement digital transformation precision. She has published extensively in the research fields of HCI and educational technology, with more than 150 peer-reviewed research papers and several educational technology books, and has been awarded numerous awards for funded projects.

ORCID profile https://orcid.org/0000-0001-7547-9616

David Towl is a Health Promotion Practitioner with particular interest in Alcohol and Drug-related harm, Sexual and Reproductive Health and Homelessness. David is currently employed as the General Manager Health Promotion and Engagement at Access Health and Community—a non-profit health service in the east of Metropolitan Melbourne.

Zac Haussegger is a research assistant who has worked at Swinburne University for the Department of Social Work over the past 3 years, most notably on public health projects examining cross cultural and health and wellbeing in relation to risky drinking. He has skills in academic writing, editing and data analysis. He is currently studying medicine at the University of Melbourne and has a keen interest in public health.

ORCID profile https://orcid.org/0000-0001-9757-0330

The Adoption of the Rasch Measurement Approach as a Learning Analytics Tool in Higher Education

Allaa Barefah

Abstract This chapter argues that the Rasch measurement approach has valuable contributions that have not yet been fully utilised in applying learning analytics to optimise the learning experience in higher education. The chapter aims to show how the application of the Rasch approach offered some practical solutions for redressing the operational challenges facing the learning analytics field. The results reported from a Ph.D. research project conducted at four higher education institutions in the Kingdom of Saudi Arabia. Data collected from 162 undergraduates voluntarily participated in a series of 2×3 quasi-experiments. The chapter presents the Rasch-driven analyses and explained its practical implications concerning three key elements in the learning process: learners, context and assessment.

Introduction

With the very limited resources available, higher education institutions adopt a data-driven decision-making approach to improve their programmes' efficiency. Although the growing field of learning analytics has informed educational leaders in leading innovations to optimise the learning experience, there remain opportunities to expand its potentials to instructors and learners. The Rasch model proved its effectiveness in educational measurement and showed considerable promise in the calibration of testing instruments for assessment. However, its application in the LA field is still underdeveloped. This chapter argues that the Rasch measurement approach holds considerable promises to provide some practical solutions fpr redressing the operational challenges that are currently facing the LA field.

The chapter showcases the application of the Rasch measurement approach, in a doctoral research project, as a LA tool to optimise the learning for HE learners in the Kingdom of Saudi Arabia (Barefah, 2019). The first section of the chapter provides

A. Barefah (✉)
Taif University, Taif, Saudi Arabia
e-mail: Aobarefah@tu.edu.sa

E. McKay (ed.), *Manage Your Own Learning Analytics*, Smart Innovation, Systems and Technologies 261, https://doi.org/10.1007/978-3-030-86316-6_6

an overview of LA and its major challenges. Research methodology was explained in the second section. The chapter continues to present the Rasch-driven analyses and discuss its practical implications in relation to three key elements in the learning process; learners, context and assessment.

Learning Analytics

Consensus does not yet exist on a precise definition for the nascent field of Learning analytics. However, most of the exiting definitions revolve around the definition suggested by Ferguson (2012) who proposed that LA is:

> *the measurement, collection, analysis and reporting of data about learners and their contexts, for purposes of understanding and optimising learning and the environments in which it occurs*

Existing works on LA vary in its purposes. Some research focuses on identifying the technological, educational or political factors driving the adoption of LA in educational contexts (Ferguson, 2012). Other studies investigate emerging trends of LA, limitations, benefits, methods and the key challenges (Avella et al., 2016; Leitner et al., 2017). Whether the focus of previous studies was on learners, instructors, or institutions, much of the conducted research in LA is conceptual and review-based (Viberg et al., 2018).

Major challenges on an operational level remain that limit the advancement of the field. The practical side of LA is very limited in use. This is because institutions deal with vast amounts of data that complicate the operationalisation process (Ellis, 2013). Another key challenge in the field of LA is the lack of clarity regarding the pedagogical relevance of LA and the real value in contextualised settings of learning across different disciplines (Tan & Koh, 2017). Additional recognised challenge is the measurement scope of learning analytics being currently narrow as it focussed on a certain group of learners, usually, those who are 'at-risk' of withdrawing or failing the course (Ellis, 2013). As described by Ellis (2013), the current measurement approach applied in LA does not offer a deep level of granularity of data that could expand the use and effectiveness of learning analytics in HE context.

The Rasch measurement approach holds considerable promises that could offer some practical solutions for redressing these operational challenges facing the LA field. This chapter shows how the application of the Rasch measurement approach as a LA tool enabled meaningful measurement and provided granular data for contextualised analytics, personalised analytics and assessment analytics.

Research Methodology

Data were collected in four research phases that consisted of three validation studies (pilot study 1, 2 and -3), followed by the main experiment. A series of quasi-experimental 2×3 multivariate design studies were conducted at four HE institutions in the Kingdom of Saudi Arabia. A total of 162 female undergraduates participated in the staged experimental procedure which was conducted during the normal two-hour tutorial classes.

The experiments were conducted in four main procedural steps as depicted in Fig. 1. Participants' registration was conducted during the first step followed by the briefing session in which participants received a short verbal explanation of the research schedule. Then participants commenced the pre-test to assess their entry skill level (or basic DFD knowledge) prior to undergoing the instructional intervention.

Participants were then allocated randomly into the instructional interventions; either F-2-F (T1), computerised (T2), or blended (T3). The third procedural step was the instructional intervention sessions during which participants received the

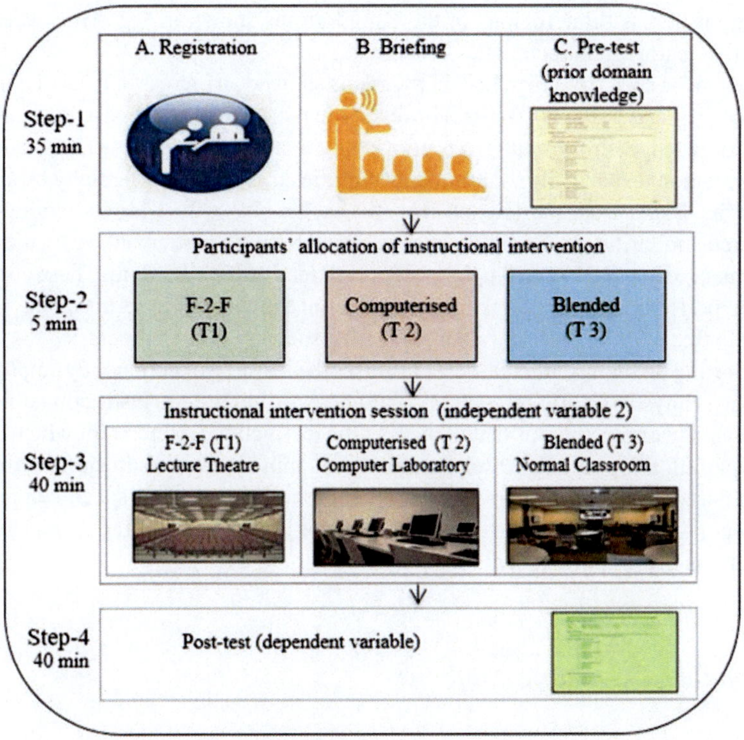

Fig. 1 Experimental procedure (Barefah, 2019 p.105)

instruction. The final procedural step of the experiment was the post-test, which was designed to assess participants' ability in terms of knowledge-change after the instructional interventions.

Results

The results of the Rasch analyses will be presented in the following section in relation to the LA key elements respectively: context, learners and assessment.

Contextualised Analytics

The Rasch analyses allowed the participating HE institutions to identify potential opportunities to offer contextualised learning that is tailored to the instructional needs of their learners. The varied performance of learners who were located at different institutions across three regions of the Kingdom (as shown in Fig. 2) revealed the distinctive learning characteristics of learners.

As it can be seen from Fig. 3, all participants showed an increased level of abilities from the pre-tests to the post-tests. However, the most obvious increase in the ability band (knowledge shifts) was from pilot study 1 and pilot 3 (the eastern and the western region respectively). Participants' abilities increased substantially by almost 1.20 logits. By contrast, participants from the capital city of the Kingdom (pilot study 2), shared similarities in the ability change estimates with participants from the main experiment who were located in the eastern province of the Kingdom. These context-specific performance variations imply unique interpretations in terms of the subtle cultural effects of the Saudi Arabian collective society on the learners.

Reporting on such contextual data optimises the learning experience by emphazing on what Tennyson (2001) refers to as 'uniqueness', i.e. unique instructional design models and learning systems that fit the specific instructional context for which it was developed. In this way, the Rasch measurement approach assists in shifting the LA design paradigm from the one-size-fits-all approach into context-specific approach that responds to the various instructional needs for learners and promotes educational innovations that impact students' learning.

Personalised Learning Analytics

The Rasch analysis provided detailed personalised analytics about how each individual learner, engaged with the instructional content, processed information and

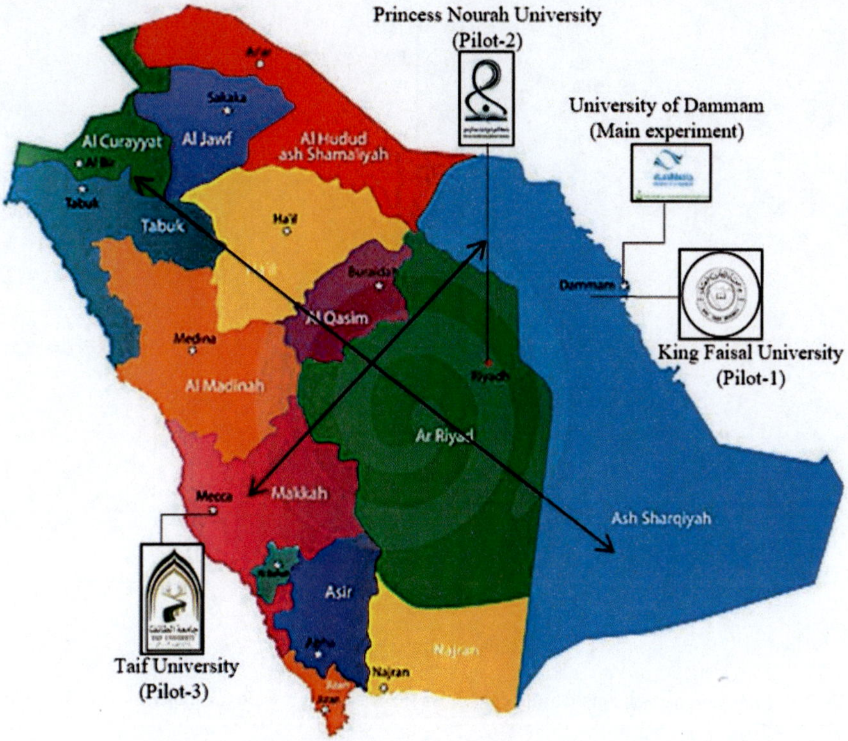

Fig. 2 Geographical locations of the participating institutions (Barefah, 2019 p. 83)

performed under three delivery modes. Most importantly, the use of the Rasch analysis allowed fair performance comparisons between and within the various groups of learners.

As learners with different cognitive style preferences (Wholist, Analytic, Verbaliser, or Imager) received their instruction through three delivery modes (F-2-F, blended and computerised), the Rasch analyses revealed details regarding the individual differences among learners. For example, the wholist learners who tend to perceive information as an overall view performed their best under the computerised mode (Fig. 4). Similarly, the best performance for the Verbaliser learners was under the computerised mode. However, the imager learners performed well under the blended mode when the instruction matched their preferences (Fig. 4). The analytic learners, on the other hand, performed their best under the F-2-F mode.

Both Wholist and Imager participants performed exceptionally well when the instructional content was tailored to their cognitive preferences. However, the unexpected performance of verbaliser and analytic learners was noted when the instruction mismatched their preferences.

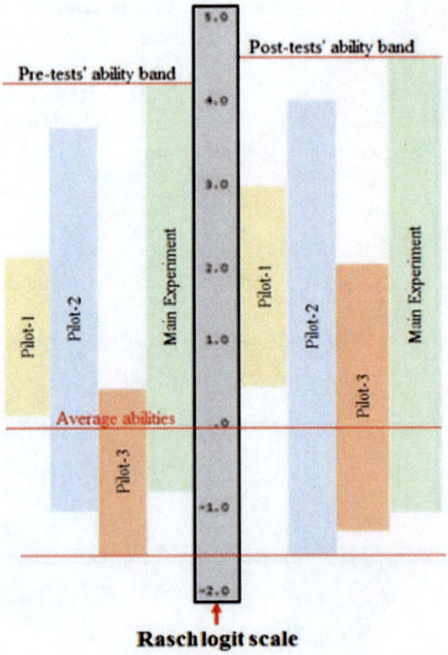

Fig. 3 Variation in participants' ability change on the Rasch logit scale (all experiments) (Barefah, 2019 p. 170)

The shared similarities and the differences in learners' performance provide instructors and courseware designers with insights to best target the digital technologies and ameliorate the ePedagogical design practices that suit individual learners. Relying on such findings, the HE institutions can optimise the learning experience in institutional data-driven decision-making processes that enhance learners' engagement and control of their learning. The results can also be interpreted to increase learners' awareness of their cognitive preferences as to make better informed decisions of how they can invest their efforts.

Assessment Analytics

The Rasch approach allows us to utilise the assessment data sets, which was considered an available asset for HE institutions that has been rarely used to inform learning analytics. The Rasch analysis did not only allow the construction of the measurement scale that reliably captured participants' performance, but also provided interpretations beyond learners' abilities. The fine-grained details informed us about learners' achievement and the problematic areas where they faced difficulties to achieve the

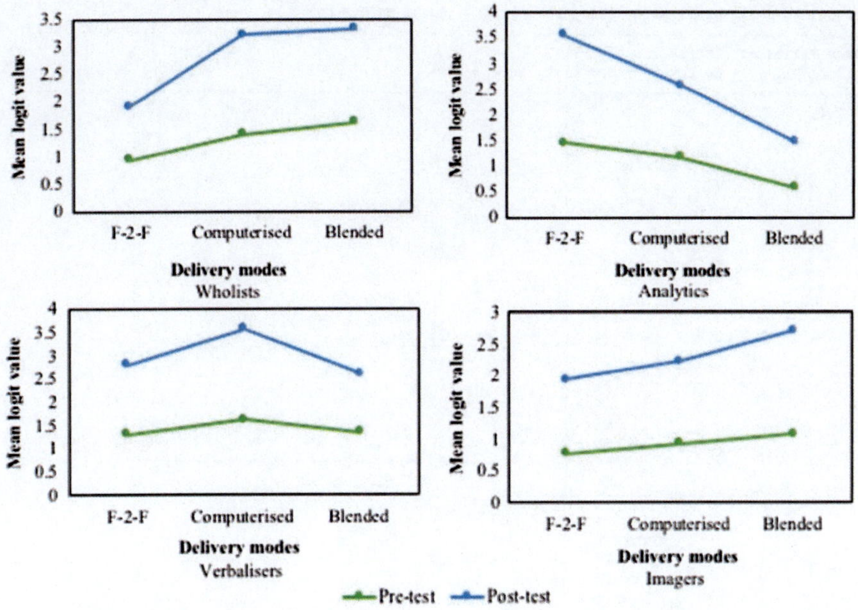

Fig. 4 Changes in mean ability estimates for wholists, analytics, verbalisers and imagers (adapted from Barefah, 2019 pp. 179–182)

required knowledge or skills. This was an important indication to evaluate the effectiveness of the ePedagogical design practices and the validity of the test items. The problematic areas identified by the Rasch analysis were further analysed and the testing instrument was revised accordingly.

The QUEST variable maps were used as a diagnostic measure to assess the overall difficulty of the test and improve its validity. For example, analysing the distribution of items' difficulties relative to participants' abilities in Fig. 5 identified the areas of knowledge domains that lacked sufficient test items. The required changes were made in an attempt to revise the testing instruments and improve their precision during the developmental refinement stages.

Also, the QUEST kid map analysis allowed us to assess the performance of each individual (participant). It was possible to measure the ability of each participant compared to other participants and identify her ability band on the logit scale. The kid map enabled us to identify both the easy and difficult items achieved by the participant as well as those items which were not achieved by the participant as expected by the Rasch model (example is depicted in Fig. 6). These findings helped with targeting the selection of the appropriate item format that would best measure the acquisition of a particular type of knowledge and improved the ICT-design practices for learners.

The iterative Rasch-driven validation of the assessment instruments provided insights on assessment analytics. It supports the claim that assessment analytics

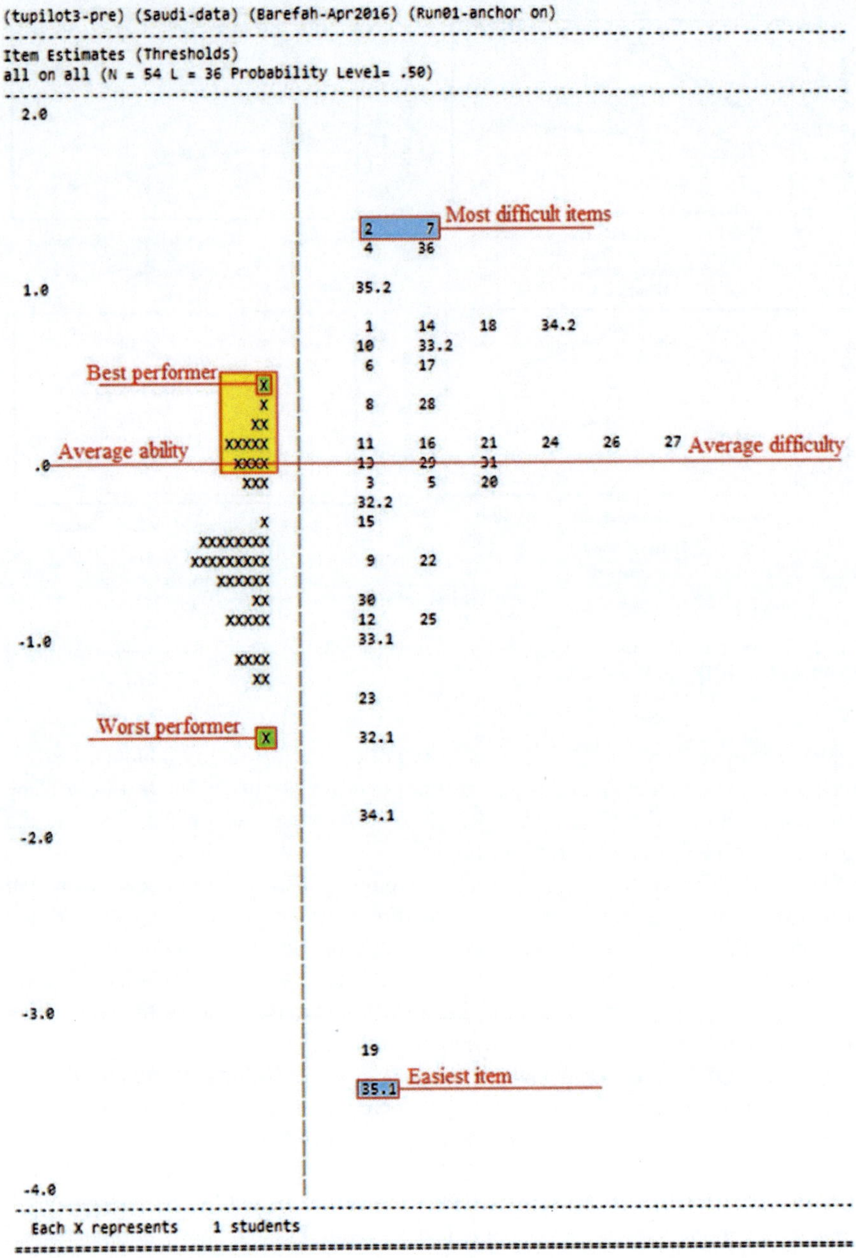

Fig. 5 QUEST variable map (Barefah, 2019 p. 143)

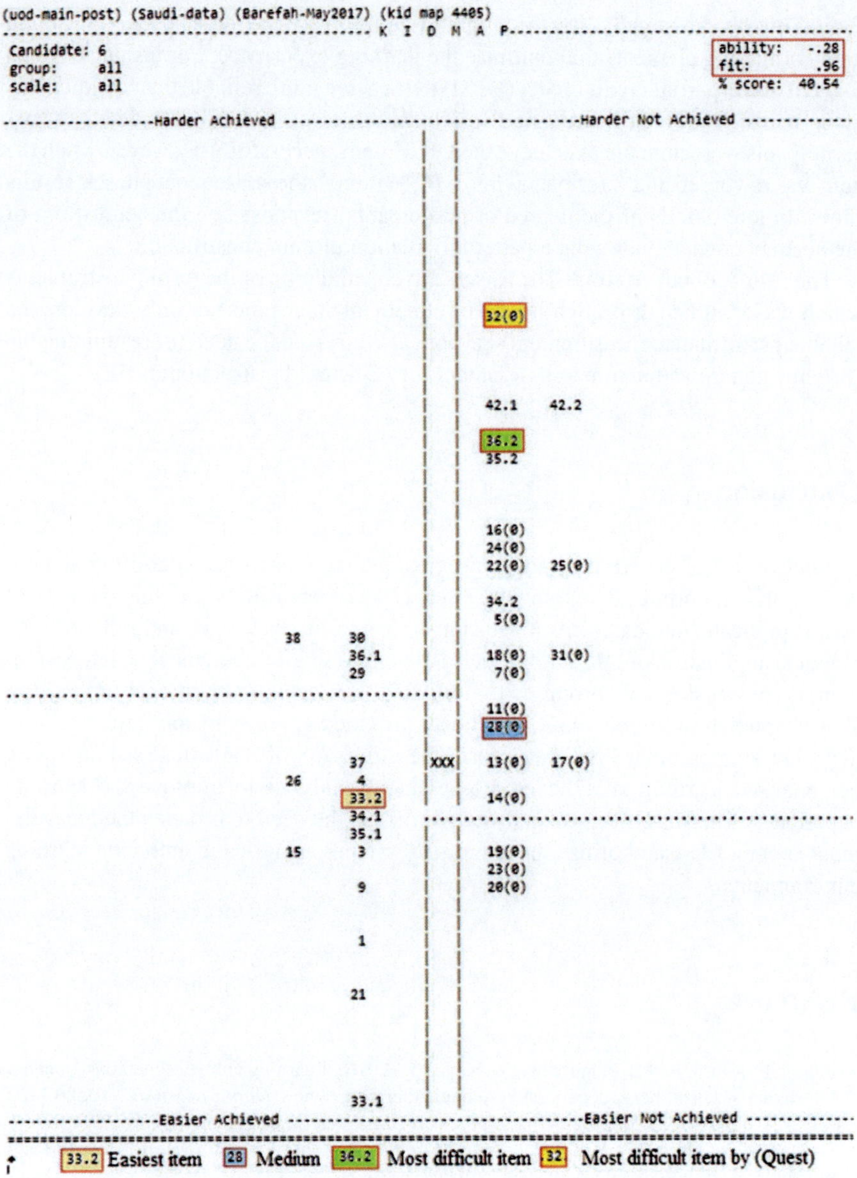

Fig. 6 QUEST kid map (Barefah, 2019 p. 191)

should not be statistically driven. It should be motivated by other pedagogical and item-formatting elements that optimise the learning experience. For instance, it was found that the partial credit model (PCM) items were more reliable than the dichotomous items (DICH) in measuring the acquisition of procedural knowledge. Many participants were not able to achieve the DICH item successfully. However, when the item was reworded and reformatted into a PCM item in the subsequent phases, results showed more details of the degree of participants' responses on the acquisition of the abstract concept, allowing a better articulation of item construction.

The results obtained from. The Rasch-driven validation of the testing instruments which was conducted through iterative refinement stages did not only developed a reliable performance measurement scale but also stressed the need for revamping the current standardised nature of assessment for ICT-based instruction in HE.

Conclusion

The field of LA offers HE institutions promising directions towards educational innovations and an optimised learning experience. The increasingly growing quantity of data complicates the data-driven decision-making process for HE and influence the practical application of LA. Within such a dynamic environment, the Rasch measurement holds considerable promises to mitigate the effects of operational limitations. This chapter showed how the Rasch model provided not only meaningful measurement but also granular data that inform the utilisation of contextualised analytics, personalised learning analytics and assessment analytics to improve the learning experience. The Rasch-driven analyses confirmed the need to broaden the measurement scope of LA and utilise the use of assessment analytics in authentic learning environments.

References

Avella, J. T., Kebritchi, M., Nunn, S. G., & Kanai, T. (2016). Learning analytics methods, benefits, and challenges in higher education: A systematic literature review. *Online Learning, 20*(2), 13–29.

Barefah, A. (2019). *An Empirical Investigation of the Effectiveness of eLearning Strategies in Higher Education: A Rasch-Model for Saudi Arabia.* Ph.D. thesis https://researchbank.rmit.edu.au/view/rmit:162687. RMIT University, Australia.

Ellis, C. (2013). Broadening the scope and increasing the usefulness of learning analytics: The case for assessment analytics. *British Journal of Educational Technology, 44*(4), 662–664.

Ferguson, R. (2012). Learning analytics: Drivers, developments and challenges. *International Journal of Technology Enhanced Learning (IJTEL), 4*(5–6), 304–317.

Leitner, P., Khalil, M., & Ebner, M. (2017). Learning analytics in higher education—a literature review. *Learning analytics: Fundaments, applications, and trends* (pp. 1–23). Springer.

Tan, J. P. L., & Koh, E. (2017). Situating learning analytics pedagogically: Towards an ecological lens. *Learning: Research and Practice*, **3**(1),1–11. 0.1080/23735082.2017.1305661.

Tennyson, R. D. (2001). Defining core competencies of an instructional technologist. *Computers in Human Behavior, 17*(4), 355–361.

Viberg, O., Hatakka, M., Bälter, O., & Mavroudi, A. (2018). The current landscape of learning analytics in higher education. *Computers in Human Behavior, 89*, 98–110.

Allaa Barefah Ph.D., is an assistant professor of Management of Information Systems at Taif University, Taif, Saudi Arabia. Dr. Barefah obtained her Ph.D. from RMIT University, Melbourne, Australia. Her research interests focus on investigating effective ePedagogical practices employed in eLearning applications in the higher education sector. Her educational technology works involve the utilisation of experimental designs and the Rasch evaluation models. Dr. Barefah is experienced in teaching Management Information Systems courses in a number of undergraduate and post-graduate programmes. She has published peer-reviewed academic articles, book chapters and conference papers.

ORCID profile https://orcid.org/0000-0002-7760-0093

Rasch Model Case Study: Investigating Cognitive Performances of School Children's Museum Learning Experiences

Asmidah Alwi

Abstract Learning analytics is used to improve learning and the environment in which it takes place. There has been much research on measuring, gathering, and analysing learner data in formal educational settings. However, its use in informal learning environments like museums appears to be promising. This chapter describes a case study investigating the cognitive performances of school children aged 10–12 years old during their museum learning experiences by considering the differences in human cognitive preferences. The research study focused on the Dinosaur Walk exhibition in Melbourne Museum as the context of the investigation. The main objective of the research study was to look into the relationship in the form of interactive effect between the students' cognitive preferences and the museum information representation formats in facilitating their museum learning experiences. This research is important as museum learning has been well accepted as informal learning to support formal education. Hence, this chapter serves as an example of managing learning analytics within the informal educational domain. The case study can be exemplary to the education sector and related organisations or businesses towards understanding, planning, and assessing students' performances and preparing a better way of facilitating/supporting learners' learning processes and experiences.

Introduction

Research on the design of multimedia learning can be approached by either focusing on the technology, which focuses on the capabilities of the advancing technologies or the learner that focuses on the human cognitive system (Mayer, 2009). The first approach is relatively straightforward, using appropriate multimedia technology to enhance the learning process, often towards finding the best multimedia approach to represent the information. While on the other hand, the second approach is centralised around the learner and seems more complicated, whereby the concentration is given

A. Alwi (✉)
School of Creative Industry Management and Performing Arts, Universiti Utara Malaysia, Sintok, Malaysia
e-mail: asmidah@uum.edu.my

E. McKay (ed.), *Manage Your Own Learning Analytics*, Smart Innovation, Systems and Technologies 261, https://doi.org/10.1007/978-3-030-86316-6_7

towards aiding human cognition. To do so requires a deep understanding of how humans think and process information (Ali et al., 2014; Riding & Rayner, 1998).

Designing and developing digital materials for complex pedagogical environments requires detailed attention creating many challenges for interactive multimedia designers, particularly when deciding how to deal with the evolving interfaces and interaction design requirements, and other human–computer interaction (HCI) issues such as technical requirements to support people's interactions with the digital environment. Hence, addressing such critical factors of human differences in HCI design is crucial to ensure the usability aspects of digital learning systems. Moreover, the diversity of learners' cognitive instructional and learning preferences forced careful and detailed consideration in addressing an individual's instructional and learning needs online. For instance, individual cognitive styles relate to HCI through the learners' preferred information processing mode on their performance on a computer and their overall preferences for specific design elements (such as their preferences towards verbal or visual information representation formats). Therefore, various instructional strategies that utilise multimedia have been adopted to enable authentic pedagogies. Instructional strategies involving multimedia thereby provide a broad range of opportunities for the design, with many new possibilities for innovative instructional/learning environments (Coombs & Bhattacharya, 2018). Initially, instructional materials' design and development process seem relatively straightforward based on the instructional objectives. However, the production of original, quality materials with appropriate visual designs that meet usability are among the challenges instructional designers must consider while maintaining their originality (Şendurur et al., 2018).

The challenges of using technology in education and its anticipated benefits in resolving the issues commonly connected with contemporary educational practices are never-ending. The idea of analysing data on students and their environments is not new in education. Long before learning analytics, educational research has been aware of many types of learning technologies and learning data. And recently, learning analytics has been a perpetual interest of many in the educational field. Learning analytics has been defined as measuring and reporting on data concerning learners and their different contexts to understand and optimise their learning and the environmental contexts (Siemens et al., 2011). With the ability to anticipate individual student learning performances at designated points, learning analytics aims to better understand and optimise learning by providing customised and timely assessment (Joksimović et al., 2019). The use of learning analytics to better understand informal educational settings such as museums is significant.

This chapter describes a case study investigating the cognitive performances of school children aged 10–12 years old during their museum learning experiences by considering the differences in human cognitive preferences. The research study focused on the Dinosaur Walk exhibition in Melbourne Museum as the context of the investigation. The main objective of the research study was to investigate the relationship in the form of interactive effect between the students' cognitive preferences and the museum information representation formats in facilitating their museum learning experiences. Hence, this chapter would serve as an example of managing learning analytics within the informal educational domain.

Study Background

Museum learning is recognised as being informal learning. Nevertheless, the design of museum learning experiences uses different learning theories (Schaller et al., 2007) to suit the desired instructional outcomes. In practice, specific attention is given to measuring the instructional outcomes and the effectiveness of the museum learning programmes (Hooper-Greenhill, 2004). Students and teachers have become a target audience for adopting museums' content (Hammerness et al., 2016) within their educational curriculum. Many countries around the globe, such as the United Kingdom, Australia, and the United States of America, have utilised museums to facilitate their formal education. Smithsonian Centre for Learning and Digital Access with the School of Education at the University of California for example, has come out with a research project to understand how the museum can best help teachers utilise the museum content in supporting their teaching. More importantly, due to the increasing demand of the web-based or online museum, equal attention needs to be given to the design of the web-based or online museum exhibits towards facilitating the experiential learning process. Therefore, a more systematic investigation to measure and provide empirical evidence of the effectiveness of web-based museum environments that facilitate successful learning experiences is needed.

Over the years, the literature has discussed various types of learning in the museum. Often, the idea of museum learning has been interchangeably used or referred to as the process of meaning-making of the museum experience (Falk & Dierking, 2000; Rennie, 2015). In contrast to formal learning, the process of museums learning resides within the visiting experiences in a free-choice setting without the requirements of a curriculum set and student attendance; thus, determined by the visitors' locus of control (Falk & Dierking, 2002).

With the latest technology, various multimedia instructions have been implemented as part of the museum exhibits. Although, recent trends show their use of virtual reality (VR), augmented reality (AR), and mixed reality (MR) as their latest additions to their exhibition design arsenal. Nevertheless, the common representation or instructional content of museum exhibits using textual and visual formats remain essential.

There are many new ways to present instructional content, which adds variety to instructional tactics and helps to create engaging new learning environments (Abu Bakar et al., 2018; Koçak et al., 2019). The notion that multimedia instruction should include both words and pictures to improve learning (Mayer, 2014, 2017) has motivated ongoing efforts to develop efficient and effective instructional events (for example see: Seufert, 2019; Mason, 2018) suggesting that such multimedia artefacts play a crucial role in an educational setting (Chan et al., 2018; Savov et al., 2019). It is important to note that, regardless of how well the instruction has been designed, individuals will respond to different forms of instruction in different ways (Price, 2004; Sandler-Smith and Riding, 1999). Hence, based on the instructional theories, instructions must be designed based on how individuals learn best (Gagne et al., 1992; Laurillard, 2006), considering the situation that accommodates the learning process (Gagne, 1985; Gagne et al., 1992).

The full potential of (educational) technologies can only be realised if the technologies' design and use are based on a thorough understanding of individual learning needs. (Koć-Januchta et al., 2019; Bower, 2019). Furthermore, Holden (2015) asserts instructional designers to think about relevant learning theories that explain how people learn differently; for example, understanding different learning styles so that appropriate methods and materials may be built and developed to help people learn better. Furthermore, multimedia literature focusing on learning and instructional design recommend that well-designed educational programmes take both the human cognitive perspective and multimedia principles into consideration (Alemdag, 2018; Merrienboer & Kester, 2005).

A person's cognitive style is a psychological dimension inextricably tied to their cognitive system (Peterson et al., 2009). Each person's cognitive style preference is distinct and likely to be a permanent feature of their cognitive functioning (Riding & Rayner, 1998; Sadler-Smith & Riding, 1999). This opinion assumes that people learn differently and that these differences are identified and quantified (McEwan & Reynolds, 2007). Additionally, cognitive styles are implicated in how individuals perceive and interact with learning environment to shape both their own brains and their learning environments. As a result, cognitive preferences defined as an individual's preferred and habitual method of organising and representing information. Cognitive preferences provide a more comprehensive and useful definition of students than intellectual abilities alone (Messick, 1984). Academics have tried to prove that cognitive preference outcomes can improve expected educational performance. Messick (1984) proposed six educational, cognitive styles that should include: (i) improving the instructional methods by providing a foundation to guide the appropriate presentation (delivery) mode; (ii) providing the opportunity to understand better students' way of thinking (information processing) which may help to broaden the educational goals and outcomes; (iii) enhancing students learning and thinking strategies; (iv) enriching teacher behaviour and conceptions; (v) expanding guidance and vocational decision making and finally (vi) tuning the stylistic demands of learning environments.

Cognitive Styles

Sternberg and Grigorenko (1997) divided cognitive styles into three groups: cognition-centred, personality-centred, and activity-centred approaches. The cognition-centred approach was further organised into three models or labels (Riding & Cheema, 1991; Rayner & Riding, 1997, 1998): (i) relate principally to the cognitive organisation (Wholist-Analytic style), (ii) relate principally to mental representation (Verbal-Imagery style), and (iii) reflect a deliberate attempt to integrate both dimensions of the Wholist-Analytic and Verbal-Imagery. Table 1 below lists the key work reflecting the cognition-centred approach.

Riding and Cheema (1991) recommended integrating both styles as significant in structuring how people receive information either as a whole or in parts, and secondly, representing information or thinking in words or pictures. Nonetheless, Sadler-Smith and Riding (1999) argued that the cognitive style preferences depicted as information processing and thinking mode were orthogonal dimensions and should treated as separate continuums.

Table 1 Descriptions of styles dimension (Riding & Rayner, 1998, pp. 20)

The wholist-analytic dimension		
Field-dependency-independency	Individual dependency on a perceptual field when analysing a structure or form which is part of the field	Witkin and Asch (1948), Witkin (1964), Witkin et al. (1971, 1977)
Levelling-sharpening	A tendency to assimilate detail rapidly and lose detail or emphasise detail and changes in new information	Klien (1954), Gardner et al. (1959)
Impulsivity-reflectiveness	Tendency for quick as against a deliberate response	Kagan et al. (1964), Kagan (1966)
Converging–diverging thinking	Narrow, focused, logical, deductive thinking rather than broad, open-ended, associational thinking to solve problems	Guildford (1967), Hudson (1966, 1968)
Holist-serialist thinking	The tendency to work through learning tasks or problem solving incrementally or globally and assimilate detail	Pask and Scott (1972), Pask (1976)
Concreate sequential/concrete random/abstract sequential/abstract random	The learner learns through experience concrete and abstraction either randomly or sequentially	Gregorc (1982)
Assimilator-explorer	Individual preferences for seeking familiarity or novelty in the process of problem solving and creativity	Kauffman (1989)
Adaptors-innovators	Adaptors prefer conventional, established procedures and innovators restructuring or new perspectives in problem spreferred modes of expressionolving	Kirton (1976, 1987)

(continued)

Table 1 (continued)

Reasoning-intuitive active-contemplative	Preference for developing understanding through reasoning and or by spontaneity or insight and learning activity which allows active participation or passive reflection	Allison and Hayes (1996)
The verbal-imagery dimension		
Abstract versus concrete thinker	Preferred level ad capacity of abstraction	Harvey et al. (1961)
Verbaliser-visualiser	The extent to which verbal or visual strategies are used to represent knowledge and thinking	Paivio (1971), Riding and Taylor (1976), Richardson (1977), Riding and Calvey (1981)
An integration of the wholist-analytic and Verbal-Imagery dimensions		
Wholist-Analytic, Verbal-Imagery	Tendency for individual to process information in parts or as a whole and think in words or pictures	Riding (1991), Riding and Cheema (1991), Riding and Rayner (1995)

Wholist-Analytic Dimension

The Wholist-Analytic dimension is intrinsic, according to Riding and Rayner (1998). This cognitive-dimension affects the individual's preference for instructional delivery method, media, and learning performance by affecting the actual organisation and structure of the information received by the individual, which is either organised as wholes or as parts (Sadler-Smith & Riding, 1999). Wholists are prone to seeing concepts as wholes and are unlikely to be able to break down information into smaller chunks. On the other hand, analytics prefers to handle data in chunks and finds it challenging to combine smaller pieces of data into a single entity. So every individual's cognitive preference is unique and thus is likely to be a fixed component of the individual's (cognitive) functioning (Riding & Rayner, 1998; Sadler-Smith & Riding, 1999). Individuals can function at their best within the Wholist-Analytic dimension if they have the proper information structure.

Verbal-Imagery Dimension

The Verbal-Imagery dimension describes an individual's manner of thinking about the information they receive and might vary depending on the activity (Riding & Rayner, 1998). Because the Verbal-Imagery dimension interacts with how the information was presented, such as in the text, photographs, and diagrams, it is expected that someone who prefers text over images will do better when given a textual information piece (Sadler-Smith & Riding, 1999). Moreover, Verbalisers may work better with verbal information, whereas imagers may work better with spatial information (Graff, 2003; Riding & Rayner, 1998). Antonietti and Giorgetti (1998), on the other hand, have questioned the notion that a person's strength is limited to a single dimension (verbal or visual). They established that the verbal and visual dimensions are independent and that a person could be strong or poor in both aspects (McEwan & Reynolds, 2007). The anatomical and functional independence of the verbal-visual dimension was also established in a neuroscience investigation (Kraemer et al., 2009). In addition, Kozhevnikov and colleagues claimed that a visual system is three-dimensional, which includes object and spatial dimensions (Kozhevnikov et al., 2005).

Individual Cognitive Preferences Construct

Individual cognitive preferences are formed from the two aspects of Wholist-Analytic and Verbal-Imagery based on this notion. As a result, a person's cognitive style is likely to fall into one of the following groups: (i) Wholist-Verbaliser, (ii) Wholist-Imager, (iii) Analytic-Verbaliser; and (iv) Analytic-Imager. Each of the four style groups may have distinct basic preferences for instruction based on observable behaviour choices (Riding & Rayner, 1998). Learners who fall into the Analytic-Verbaliser type, for example, may prefer text over those who fall into the Analytic-Imager category, who may perform better with a captioned picture or diagram (McKay, 2000). Figure 1 illustrates this characteristic.

Since the case study measured participants' cognitive preferences, the researcher used the Riding (1991) cognitive style analysis tool (CSA) when developing the research instrumentation. The CSA screening test had 40 items for the Wholist-

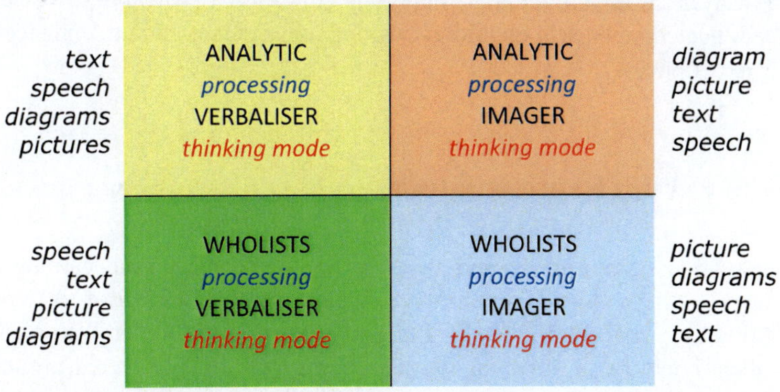

Fig. 1 Possible preferred modes of expression (Alwi, 2020, pp. 23),cognitive preference style (Riding & Rayner, 1998, pp. 156)

Analytic dimension and 48 items for the Verbal-Imagery dimension. The test was appropriate for various age groups and abilities and used in different educational settings and cultures (Riding, 2003).

Multimedia Learning

Multimedia learning strategies intended to address three distinct pedagogical objectives while supporting three different learning stages: information acquisition, response strengthening, and knowledge construction (Mayer, 2009). The information acquisition view of learning describes the process of adding information to a learner's memory. The learners are viewed as passive in this perspective, only receiving instructional information from the multimedia. As a result, concerns like what the learner gets from the learning process when the multimedia presentation aims to create a solid mental framework arise. On the other hand, the response strengthening view conceives multimedia as a reinforcer, assisting students in reinforcing what they already know. The knowledge construction approach considers multimedia learning a sense-making activity in which the learner must actively design their mental model based on the multimedia presentation. It became more personal as the student is responsible for their learning, with teacher support and multimedia presentation as tools, emphasising the critical role of multimedia in helping the learning process.

Nevertheless, in a bigger picture, it can be assumed that the earlier stage (information acquisition) can gradually move to another (finally to knowledge construction) as people build their mental model. More importantly, the expected learning outcome depends highly on the integration of all the learning stages. Knowledge retention, or the ability to remember learning material presented in the same way after some time, and knowledge transfers, or the ability to solve problems using the knowledge learned (Mayer, 2002), are the most important educational goals. For instance, this could be reflected when schoolchildren visiting a museum exhibit to enhance their learning experiences.

Museum Exhibit Representations

The employment of multimedia and digital technology enhances the learning experience by providing a more comprehensive range of representation formats (Holewa, 2017; Middleton & Cockrill, 2017). Presentations in an online or Web-mediated environment could be in many modes, allowing instructions delivered in multiple ways. As such, the diverse approaches used in the design of online museum exhibits reflect the implied acknowledgement of individual cognitive variations, hoping that learners will learn meaningfully. Although there are various media alternatives to depict museum exhibits in accommodating multiple modalities, the museum mainly uses verbal (text) and visual (pictures) in their exhibit presentation strategies. Because multimedia instructions rely significantly on a well-designed environment to suit the diverse levels of intellectual ability of distinct museum visitors, a museum exhibit must be linked with the proper instructional methodologies (McKay et al., 2003) and effectively blended (Deshpande et al., 2007).

This fundamental concept of instructional design of the museum exhibition, the different human cognitive preferences and the museum information representation formats built up the research framework of the case study. This case study utilised the benefits of learning analytics in conducting a systematic investigation measuring the effectiveness of online museum environments to facilitate a successful learning experience by considering the students' cognitive preferences.

Research Implementation

This case study employed a quantitative research approach to correctly capture an individual's information processing capacities by determining the students' cognitive media preferences and prior domain knowledge relevant to the Dinosaur Walk exhibition as the chosen museum instructional concept. The research was conducted within the museum web-based instructional format (Treatment 1) and the physical museum environment (Treatment 2). Overall, there were three phases involved in the research implementation: (i) Phase1—Assessment instrument construction (development of the pretest and posttest items questionnaire), (ii) Phase 2—Calibration study (for the validity and reliability testing of the pretest and posttest instruments and the investigative research procedure) and (iii) Phase 3—The main experiment.

The assessment instrument construction in Phase 1 was carefully planned. To ensure that the instruments were appropriate for the participants' age group, the design and development of the measurement instruments considered the Victorian Essential Learning Standard (VELS) framework, which was adapted in the Dinosaur Walk Exhibition to guide the content development process. The Department of Education and Early Childhood Development (DEECD) of Victoria, Australia, derived the VELS to outline important learning aspects for all Victorian students from Prep to Year 10. VELS provided the standard guideline for all Victorian schools to plan student learning and assessment and prepare a report to parents.

The research procedure conducted in Phase 2 and Phase 3 employed a three-phase quasi-experimental methodology. The first phase of the research procedure involved the CSA test, whereby all participants were required to take this test. Based on the cognitive style ratios generated from the test, the participants were randomly assigned into the treatment groups which have been visualised in Fig. 2.

Before the museum visiting treatment, the pretest was administered. The post-test was conducted after the museum-visit treatment (online or physical). This methodology allowed the observation of participants' performance outcomes based on the treatments and, at the same time, offered the flexibility to explore the differences in participants' performance outcomes between both treatments. Figure 3 depicts the overall research phases.

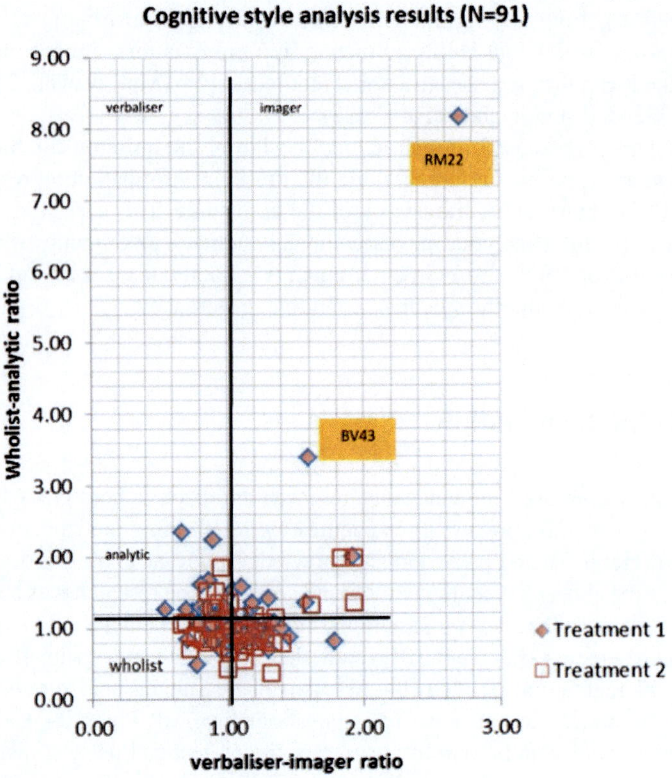

Fig. 2 Scatter graph of the CSA ratio ($N = 91$) (Alwi, 2020, pp. 160)

Rasch Analysis as the Analytic Learning Tool

The capacity to test the participants' cognitive performance in connection to the learning outcomes of their museum visit was crucial to this case study. As mentioned earlier, a pretest and a posttest were included as a set of evaluation instruments that were used to track participants' progress in the Dinosaur Walk exhibition learning experiences. The goal of the assessment instruments was to show that all test-items measured the same construct and show that the test-items can effectively distinguish between those who lack knowledge and those who are knowledgeable about the Dinosaur Walk museum's exhibit information.

The Rasch measurement model was appropriate to analyse an individual's performance of the participant's relative performance concerning the test-items difficulty on the same unidimensional measurement scale. Based on the analysis, it should be expected that the participant with high ability will get the easiest item correct, whilst the participant with low ability to get the most difficult item wrong. This analysis, therefore, provided an opportunity to examine closely each participant performance based on the test-items. For example, Fig. 4 (the person-item/variable map)

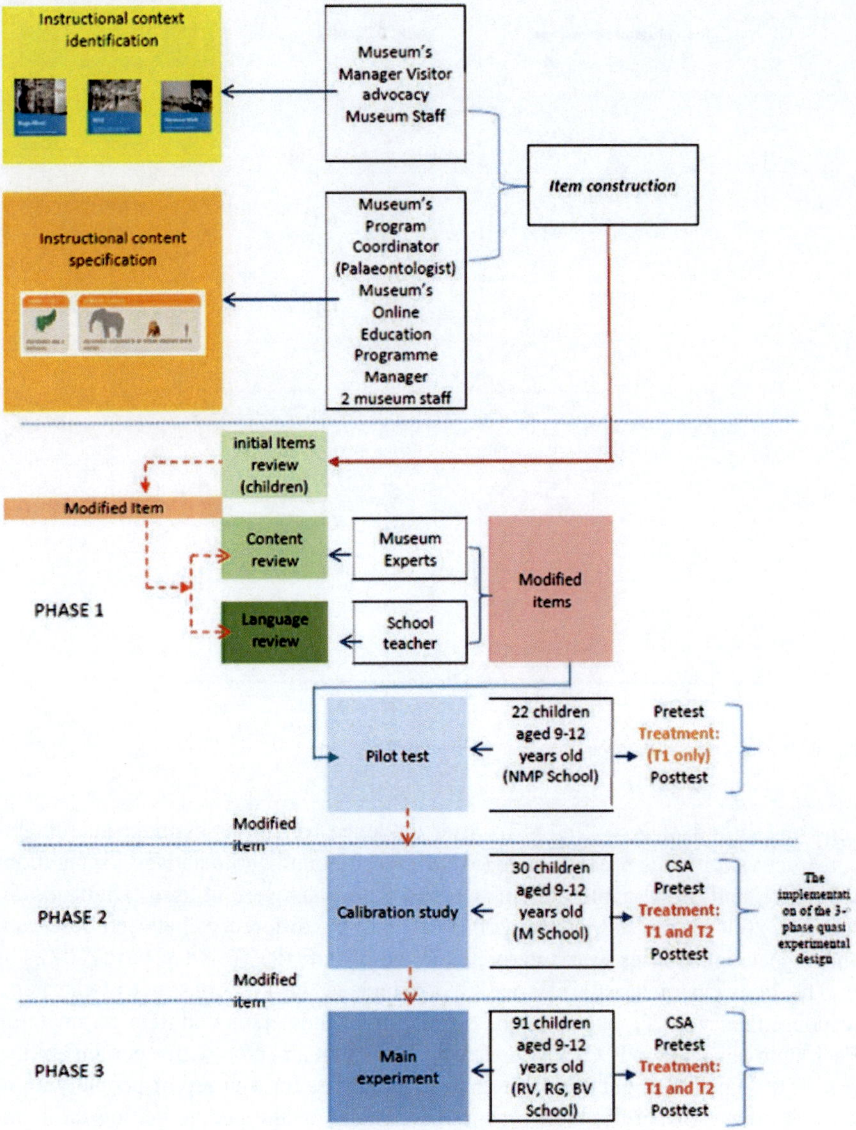

Fig. 3 The overall research phases (Adapted from Alwi, 2020, pp. 83)

visualised the individual performances over the item difficulty, which would effectively analyse the participants' learning performances. The Logit scale linearised the relationship between persons and item difficulty, where the probability of success depends on the difference between the ability of the persons and the difficulty of the item (Bond & Fox, 2007). Furthermore, the unidimensionality addressed by Rasch analysis is necessary to ensure that all test-items reflect the same underlying

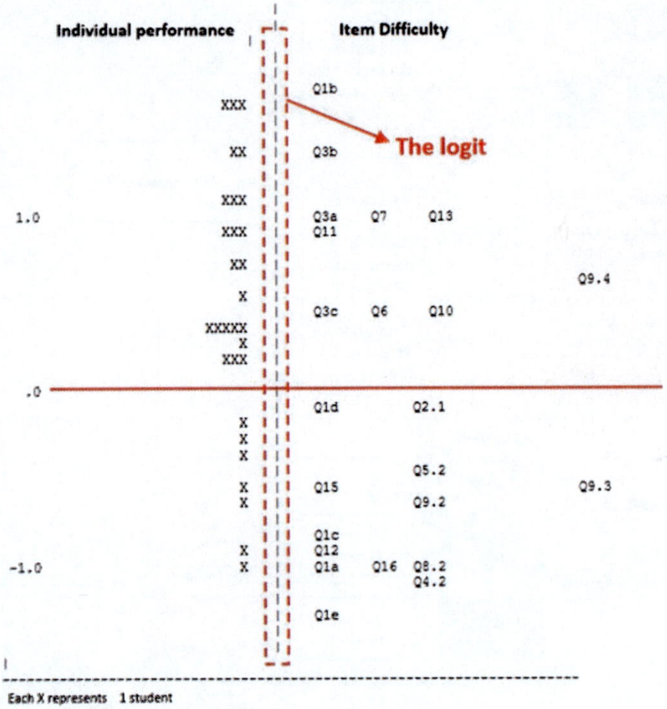

Fig. 4 Person-Item map or Variable Map (Alwi, 2020, pp. 78)

construct (Bond & Fox, 2007). To offer evidence for the assessment instrument construct validity, the Mean Square (MNSQ) of infit and outfit statistics, separation reliability, and the variable map of the Rasch analysis were utilised (Draugalis & Jackson, 2004; Ryan & McCrae, 2005). In contrast, differences between observed and expected responses were interpreted in the fit statistics (Bond & Fox, 2007).

The Item Characteristic Curve (ICC), which is the key construct of the item response theory (IRT) visualised the estimation of the person and item parameters. Each item has its own ICC, which clearly shows that a person with a certain ability has a 50% chance to get a correct answer. Whereas, for a group of people with a different range of ability, there was a 50% chance of the people getting the item correct (Wu & Adams, 2007). The ICC reflects values between 0 and 1. Woods and Baker (1985) suggested that IRT model assumes, despite how able the person may be, the probability for them to give a correct response is never certain, therefore, the model fit statistics were generated based on the difference between observed and expected responses enabling detail diagnostic ability. Figure 5 illustrates the utilisation of IRT, Rasch model and the individual cognitive preferences in the case study.

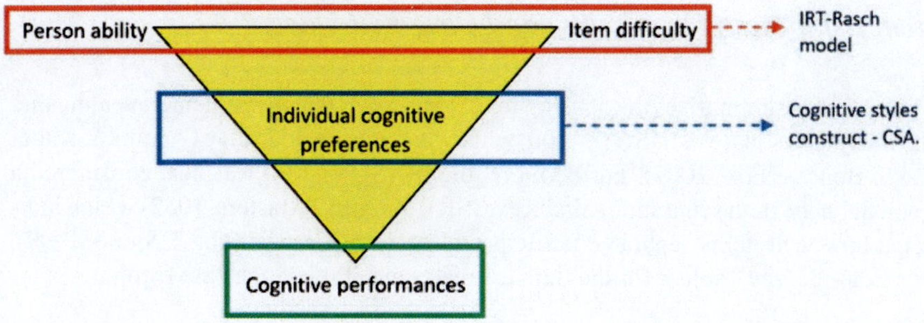

Fig. 5 Utilisation of IRT—Rasch model and individual cognitive preferences in determining cognitive performance (Alwi, 2020, pp. 9)

Data Analysis

The Rasch analysis was conducted using Quest Interactive Test Analysis System (Adams & Khoo, 1996). The test-items were reviewed based on these Quest estimates to determine (i) how successfully the test-items discriminated across participants; (ii) whether the response categories were scored in the correct order; and (iii) whether the test-items were assessing the same construct (Wu & Adams, 2007). This analysis was an iterative process. First, the inappropriate test-items were identified and then appropriate action towards those test-items was taken based on the elimination criteria set by the Quest fit statistics. It was also important to note that the test-items difficulty values were established based on the calibration study, hence provided the anchored values for further investigation in the main experiment.

The analysis was derived based on the following Quest outputs:

Variable Map

Figure 4 shown above is the variable map that shows the relationship between test-item difficulty and the participants' performance.

Fit Map

The item fit map visually represents the item fit. In Fig. 6, −0.5 to 2.0 (horizontal) represents the Infit MNSQ, by which the dotted vertical lines represent the acceptable value of the Infit MNSQ between 0.70 and 1.30. The asterisks indicate the magnitude of the fit statistic for the item on the same line. Thus, those lie within the dotted lines are considered acceptable (Adams & Khoo, 1996). Figure 6 shows the example of the item fit map.

Item Estimates (Thresholds) and Case Estimates

This estimate summarises the fit statistics of the assessment instrument, revealing that the Infit mean square (MNSQ) is 1.00, which fits the expected value (Adams & Khoo, 1996; Bond & Fox, 2007). Furthermore, the MNSQ of 1.00 was also considered a good fit on the item separation reliability (ISR) (Wright & Masters, 1982) which indicated how well items separated participants' performance (Wright & Stone, 1999). For example, see Table 2 for the item estimates and Table 3 the case estimates.

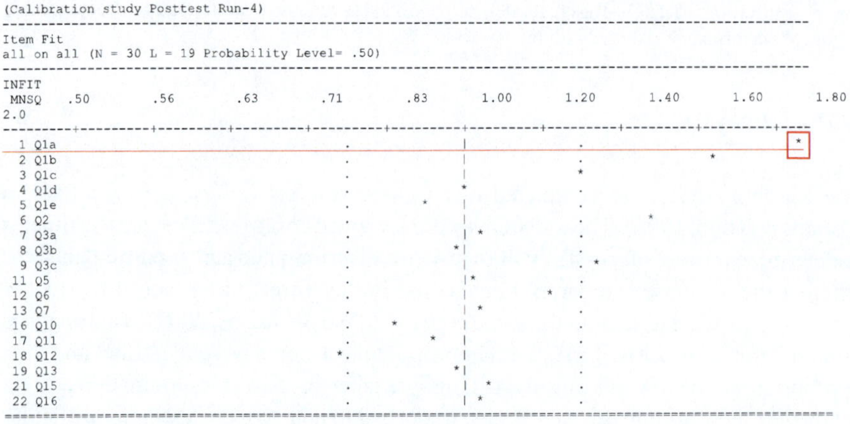

Fig. 6 Example of Quest item fit map (Alwi, 2020, pp. 137)

Table 2 Example of quest test-item estimate (Alwi, 2020, pp. 143)

Summary of item Estimates			
Mean	.14		
SD	.90		
SD (adjusted)	.90		
Reliability of estimate	1.00		
Fit Statistics			
Infit Mean Square		Outfit Mean Square	
Mean	1.02	Mean	.96
SD	.14	SD	.18
Infit t		Outfit t	
Mean	.16	Mean	.03
SD	.81	SD	.59

Table 3 Example of quest case estimate (Alwi, 2020, pp. 131)

```
Summary of case Estimates
=========================

Mean                                  .47
SD                                    .73
SD (adjusted)                         .58
Reliability of estimate               .63

 Fit Statistics
===============

 Infit Mean Square            Outfit Mean Square
    Mean        .99              Mean      1.00
    SD          .29              SD         .32


      Infit t                      Outfit t
    Mean        .04              Mean       .07
    SD          .69              SD         .68

   0 cases with zero scores
   0 cases with perfect scores
```

Item Analysis Result

The QUEST item analysis results for observed responses (Table 4) presents item performances individually. The reports enabled a closer and detailed evaluation of the test items by distributing score points and the percentage of participants within those score points. Despite that, all items were well fitted within the two standard deviations of the Infit MNSQ.

Table 4 Example of quest item analysis results for observed responses (Alwi, 2020, pp. 128)

```
Item    15: Q9                                   Infit MNSQ = 1.00
                                                       Disc =   .66

Categories                  0         1         2         3        4         5      missing

Count               0       1         3         1        12        0        13         0
Percent (%)        .0     3.3      10.0       3.3      40.0       .0      43.3
Pt-Biserial        NA    -.37      -.12      -.15      -.30       NA       .56
p-value            NA    .021      .257      .207      .053       NA      .001
Mean Ability       NA    -.92       .09      -.16       .14       NA      1.02        NA

Step Labels                 1         2         3         4        5

Thresholds              -1.72      -.66      -.52       .64      .64
Error                    1.13       .82       .79       .42      .42
```

Results and Findings

The cognitive styles analysis results were conducted based on the single cognitive style dimension (SCSD) and combined cognitive style dimension (CCSD). The results were derived from 91 students from 3 schools participating in the main experiment. As shown in Table 5, the number of participants for each CCSD is quite unbalanced (considering the participants coming from 3 different schools). Nevertheless, it should be noted that the allocation of participants into the treatment groups was random.

The Overall Performances According to Treatment Group

In general, the analysis showed that the participants in both treatments scored higher in the posttest with Treatment 2 (0.76) scored higher than with Treatment1 (0.66). However, the Treatment 2 group had higher prior domain knowledge scores during the pretest (0.85 compared to 0.34); hence, the overall performances of participants in Treatment 1 were considered better by having more improvement (+0.32). Instead, participants' performances in Treatment 2 declined with a mean difference of -0.09 between pretest and post-test. (See Fig. 7).

Although the analysis has indicated that the participants in the physical museum-visit instructional strategy (Treatment 2) had better prior knowledge, their performances seemed to decline after the treatment. In contrast, the participants of Treatment 1 (who were less knowledgeable) demonstrate an improvement after the treatment. From these results, it can be generalised that the web-based museum-visit

Table 5 Summary of participants based on cognitive preference style and distribution according to treatment (Alwi, 2020, pp. 161)

SCSD	Treatment 1 (Web-based) (N = 47)	Treatment 2 (Physical) (N = 44)	Total (N = 91)
Wholist	29	36	65
Analytic	18	8	26
Verbaliser	23	21	44
Imager	24	23	47
CCSD	*T1 (Web-based)*	*T2 (Physical)*	*Total*
Wholist-verbaliser	12	16	**28**
Wholist-Imager	17	20	**37**
Analytic-Verbaliser	11	5	**16**
Analytic-Imager	7	3	**10**

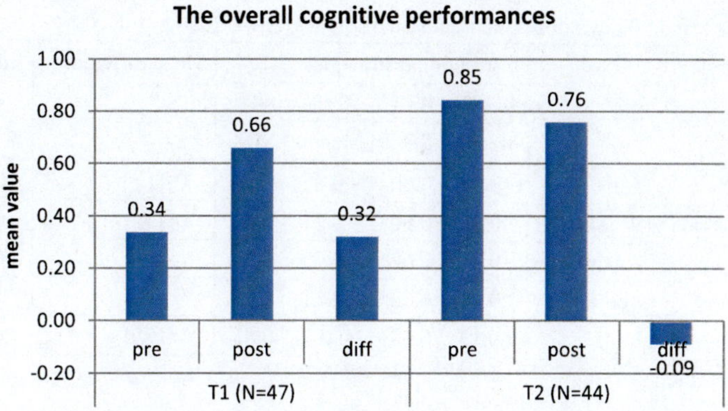

Fig. 7 The mean analysis of pretest and posttest for both treatments (Alwi, 2020, pp. 173).

instructional strategy's overall performances were better than the physical museum-visit instructional strategy. This finding hence provided evidence that web-based instruction could help to improve students learning performances.

Single Cognitive Style Dimension (SCSD) Performances

The performance of SCSD examined individually used both the Wholist-Analytic and Verbalizer-Imager continuum. The Wholist-Analytic continuum depicted a person's processing mode, whereas the Verbaliser-Imager continuum depicts a person's thinking mode (Riding et. al., 2003). This result was critical because each continuum considered a different cognitive mode, and the dimensions are independent of one another. Thus, a position on one does not affect the position on the other (Riding et al., 2003). Therefore, it is crucial to look at the result based on the different continuum rather than comparing each style individually. The analysis of SCSD is tabulated in Table 6.

Table 6 SCSD cognitive performances according to treatment (Alwi, 2020, pp. 175)

Cognitive style		Instructional strategies						Mean Diff
		T1 ($N = 47$)			T2 ($N = 44$)			
		Pre	Post	Diff	Pre	Post	Diff	
	All	0.34	0.66	0.32	0.85	0.76	−0.09	0.41
Processing mode	Wholist	0.33	0.65	0.32	0.84	0.73	−0.11	0.43
	Analytic	0.95	0.67	−0.28	0.85	0.89	0.05	−0.32
Thinking mode	Verbaliser	0.26	0.72	0.47	0.91	0.80	−0.11	0.58
	Imager	0.48	0.60	0.13	0.82	0.75	−0.07	0.19

Table 7 The standard deviation and effect size of SCSD (Alwi, 2020, pp. 175)

Cognitive style		Instructional strategies				Mean difference (T1-T2)	Effect size
		T1 ($N = 47$)		T2 ($N = 44$)			
		Mean	SD	Mean	SD		
	All	0.32	0.60	−0.09	0.74	0.41	0.61
Processing mode	Wholist	0.32	0.60	−0.11	0.69	0.43	0.68
	Analytic	−0.28	0.62	0.05	0.97	−0.33	−0.40
Thinking mode	Verbaliser	0.47	0.52	−0.11	0.61	0.58	1.01
	Imager	0.13	0.64	−0.07	0.85	0.19	0.26

As presented in Table 6, the analysis of the Wholist-Analytic continuum showed contradictory results whereby Wholists seem to perform well in Treatment 1 with mean difference of 0.32 compared to Analytics (−0.28). However, the situation was reversed in Treatment 2 in which Analytics perform slightly better with a small mean difference of 0.05 compared to −0.11 for Wholist. On the Verbaliser-Imager continuum, Verbalisers seem to benefit most in Treatment 1 with mean difference of 0.47 compared to Imager with mean diff = 0.13. Both Verbalisers and Imagers were found to have negative mean difference of 0.11 and 0.07, respectively, in Treatment 2. Further analysis on the effect size (Table 7) revealed the largest effect size was recorded by Verbaliser ($d = 1.01$), and for Wholist, the effect size was considered medium with $d = 0.68$. Meanwhile, Analytic ($d = -0.40$) and Imager ($d = 0.26$) were both considered to have small effect size.

Combine Cognitive Style Dimension (CCSD) Performances

The general results of the CCSD (Wholist-Verbaliser, Wholist-Imager, Analytic-Verbaliser, and Analytic-Imager) performances indicated that Treatment 1 was better than Treatment 2 for all CCSD. Wholist-Verbaliser was the best performer compared to the other CCSD with a mean difference of 0.67 in Treatment 1, and Wholist-Imager performed the worst in Treatment 1 with a mean difference of 0.08. In Treatment 2, Analytic-Verbaliser was the best performer with only 0.11 differences in their mean metric. However, it should be noted that the Analytic-Verbaliser group was the only CCSD that recorded improvement (with positive mean differences) compared to the other CCSD groups with declined performances (negative mean difference) after the instructional treatment. Comparison of CCSD performance in both Treatment 1 and Treatment 2 showed interesting results with Wholist-Verbaliser as the highest achiever (mean diff = 0.67) in Treatment 1 but performed the worst in Treatment 2 (mean diff = −0.18). The result also showed that two CCSD; Analytic-Imager and Wholist–Imager achievement was at par with mean difference of -0.07 for Treatment 2. The detailed performances of CCSD with instructional treatment (Treatment 1 and Treatment 2) are presented in Table 8.

Table 8 Performances of CCSD according to treatment (Alwi, 2020, pp. 177)

Cognitive style	Instructional strategies						Mean difference (T1–T2)
	T1 ($N = 47$)			T2 ($N = 44$)			
	Pre	Post	Diff	Pre	Post	Diff	
WV	0.00	0.66	0.67	0.88	0.70	−0.18	0.84
WI	0.57	0.65	0.08	0.82	0.75	−0.07	0.15
AV	0.54	0.78	0.25	1.01	1.09	0.08	0.16
AI	0.06	0.50	0.44	0.80	0.73	−0.07	0.51

Table 9 Performances of CCSD (Alwi, 2020, pp. 177)

Cognitive style	Instructional strategies				Effect size cohen's *d*
	T1 ($N = 47$)		T2 ($N = 44$)		
	Mean	SD	Mean	SD	
WV	0.67	0.48	−0.18	0.68	1.44
WI	0.08	0.56	−0.07	0.72	0.23
AV	0.25	0.50	0.11	0.31	0.33
AI	0.44	0.79	−0.07	1.76	0.37

Further analysis on the effect size is presented in Table 9. Looking at the magnitude of the effect size determined by Cohen's *d*, the Wholist-Verbaliser was determined to have a large practical and important difference (Treatment 1 mean diff = 0.67, Treatment 2 mean diff = −0.18, $d = 1.44$). The other three groups: Wholist-Imager, Analytic-Imager, and Analytic-Verbaliser were calculated to be in the small effect size ($d = 0.37$, 0.33, and 0.23, respectively).

The Interactive Effects

As for the interactive effects, Figs. 8 and 9 clearly illustrate interactive effects between the cognitive preference and instructional strategy in determining the cognitive performance outcomes of the museum visits' learning experience.

Conclusion

The purpose of the case study was to understand better children's cognitive preferences towards the museum multimedia instructional format in the museum learning experiences. The case study provides designers with a predictable measure of the instructional performance outcomes. Although this research was conducted within

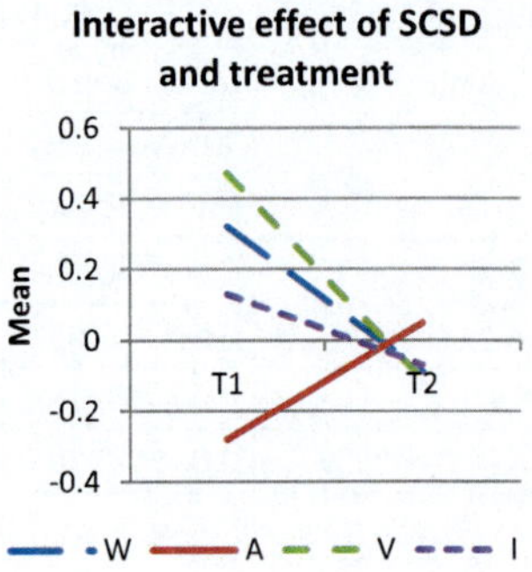

Fig. 8 The interactive effects of the SCSD (Alwi, 2020, pp. 178)

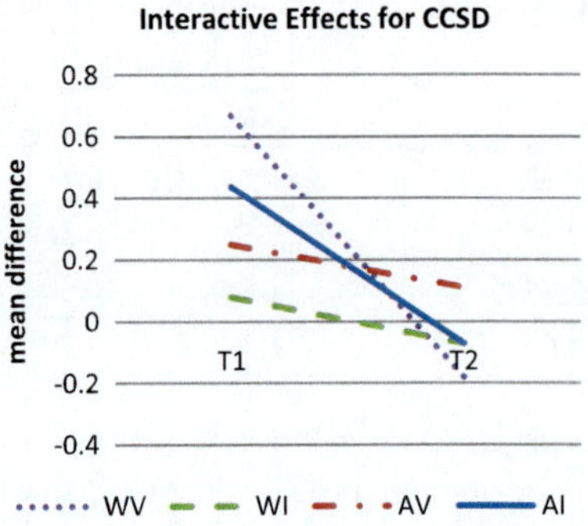

Fig. 9 Interactive effect of CCSD and instructional treatment (Alwi, 2020, pp. 179)

a specific context of a museum exhibition visit (the Dinosaur Walk in Melbourne Museum), the instructional performance outcomes inform important understandings of the interactive relationships between the design of museum-visits' instructional strategies and individual cognitive preferences towards enhancing museum-visitation learning experiences; thus, help museum exhibit curators to better understand the needs of their diverse audience (Callanan, 2012). More importantly, the research outcomes provide a quality data analysis to the research communities' body of knowledge when investigating the interactive effects of individual cognitive preferences and instructional strategies within an informal learning environment. By examining both instructional strategies (the web-based museum-visit environments and the physical museum-visit), the lessons learnt from these research outcomes can be utilised by others in designing high-quality multimedia learning environments and enhancing the design of personalised digital learning experiences.

Using the Rasch measurement model forms anchors of cognitive performance measurement and communicates aspects of student achievement (Bailes & Nandakumar, 2020). The Rasch modelling techniques offer an empirical test of construct validity while still using both item and person scores. This case study also demonstrated how the Rasch measurement model gauges the usefulness of the testing instruments. Employing Rasch measurement techniques is one way to ensure that analyses reflect the underlying data realities and, in turn, apply to fulfilling the empirical research's verifiable needs. In conclusion, Rasch modelling is a powerful approach to manage reliable learning analytics.

References

Abu Bakar, J. A., Gopalan, V., Zulkifli, A. N. & Alwi, A. (2018). Design and development of mobile augmented reality for physics experiment. In N. Abdullah, W. Wan Adnan, M. Foth (Eds.), *User science and engineering—5th international conference, i-USEr 2018, Communications in Computer and Information Science*, Vol. 886. Springer.

Adams, R. J. & Khoo, S. (1996). *Quest: The interactive test analysis system.* Australian Council for Educational Research.

Alemdag, E. & Cagiltay, K. (2018). A systematic review of eye tracking research on multimedia learning. *Computers and Education, 125*, 413–428. https://doi.org/10.1016/J.Compedu.2018.06.023.

Ali, R., Abu Bakar, Z., & Akhtar, N. (2014). The demarcation of cognitive and learning style: Myth or reality as an impediment in educational research. *Journal of Psychological and Educational Research, 22*(20), 76–101.

Alwi, A. (2020). *Interactive Effects of Information Systems Interfaces (ISI) and Personal Cognitive Preferences in Museum Learning Experiences.* http://researchbank.rmit.edu.au/eserv/rmit:163034/Alwi.pdf. (Doctor of Philosophy Full fulfilment). RMIT University.

Antonietti, A., & Giorgetti, M. (1998). The Verbalizer-visualizer questionnaire: A review. *Perceptual and Motor Skills, 86*, 227–239.

Bailes, L. P. & Nandakumar, R. (2020). Get the most from your survey: An application of rasch analysis for education leaders. *International Journal of Education Policy & Leadership, 16*(2). https://doi.org/10.22230/ijepl.2020v16n2a857.

Boldyreva. (2018). Dragicevic & Bagarić, 2019; Kärkkäinen, 2021.

Bond, T. G., & Fox, C. M. (2007). *Applying the Rasch model: Fundamental measurement in the human sciences.* Lawrence Erlbaum Associates Publishers.

Bower, M. (2019). Technology-mediated learning theory. *British Journal of Educational Technology*, 1–14, https://doi.org/10.1111/bjet.12771.

Callanan, M. A. (2012). Conducting cognitive developmental research in museums: Theoretical issues and practical considerations. *Journal of Cognition and Development, 13*(2), 137–151. https://doi.org/10.1080/15248372.2012.666730.

Chan, T. K. S., Wong, S. W. L., Wong, A. M. –Y. & Leung, V. W. –H. (2018). The influence of presentation format of story on narrative production in Chinese children learning English-as-a-second-language: A comparison between graphic novel, illustration book and text. *Journal of Psycholinguistic Research, 48*(1), 221–242. https://doi.org/10.1007/s10936-018-9600-9.

Coombs, S. & Bhattacharya, M. (2018). Smart learning requires smart thinking: The evolution of sustainable learning environments. *Journal of Educational Multimedia and Hypermedia, 27*(4), 453–470. Waynesville, USA: Association for the Advancement of Computing in Education (AACE), https://www.learntechlib.org/primary/p/181345/.

Deshpande, S., Geber, K. & Timpson, C. (2007). Engaged dialogism in virtual space: An exploration of research strategies for virtual museum. In F. Cameron & S. Kenderdine (Eds.), *Theorising digital cultural heritage.* The MIT Press.

Dragicevic, M. & Bagarić, A. (2019). Virtual technology in museums and art galleries business practice—The empirical research. In D. Tipurić & D. Hruška (Eds.) *7th international OFEL conference on governance, management and entrepreneurship: Embracing diversity in organisations. April 5th–6th, 2019, Dubrovnik, Croatia.* pp. 175–183. Governance Research and Development Centre (CIRU).

Draugalis, J. R., & Jackson, T. R. (2004). Objective curricular evaluation: Applying the Rasch model to a cumulative examination. *American Journal of Pharmaceutical Education, 68*, 1–12.

Falk, J. H. & Dierking, L. D. (2002). *Lesson without limit: How free choice learning is transforming education*. Alta Mira Press.

Falk, J. H., & Dierking, L. D. (2000). *Learning from museums: Visitors experiences and the making of meaning*. Altamira Press.

Gagne, R. M. (1985). *The conditions of learning and theory of instruction*. United States, Holt, Rinehart And Winston Inc.

Gagne, R. M., Briggs, L. J., & Wager, W. W. (1992). *Principles of instructional design*. Florida, Harcourt Brace Jovanovich College Publishers.

Graff, M. (2003). Learning from web-based instructional systems and cognitive style. *British Journal of Educational Technology, 34*, 407–418.

Hammerness, K., Macpherson, A., & Gupta, P. (2016). Developing a research agenda aimed at understanding the teaching and learning of science at a natural history museum. *Curator the Museum Journal, 59*(4), 353–367.

Holden, J. T. (2015). *An introduction to the addie instructional systems design model*. White Paper. https://static.secure.website/wscfus/4899501/uploads/white_paper--introduction_to_the_addie_isd_model.pdf.

Holewa, H. (2017). Museum futures point to collaboration on big data, artificial intelligence and digitisation. *Museums Galleries Australia Magazine, 26*(1), 12–15. Spring-Summer 2017.

Hooper-Greenhill, E. (2004). Measuring learning outcomes in museums, archives and libraries: The learning impact research project (LIRP). *International Journal of Heritage Studies, 10*, 151–174.

Joksimović, S., Kovanović, V., & Dawson, S. (2019). The journey of learning analytics. *HERDSA Review of Higher Education, 6*, 37–63.

Koçak, Ö., Yilmaz, R., Küçük, S. & Göktaş, Y. (2019). The educational potential of augmented reality technology: experiences of instructional designers and practitioners. *Journal of Education and Future, 15*, 17–36. https://doi.org/10.30786/jef.396286.

Koć-Januchta, M. M., Höffler, T. N., Eckhardt, M., Leutner, D. (2019). Does modality play a role? Visual-verbal cognitive style and multimedia learning. *Journal of Computer Assisted Learning*, 1–11. https://doi.org/10.1111/jcal.12381.

Kozhevnikov, M., Kosslyn, S., & Shephard, J. (2005). Spatial versus object visualizers: A new characterization of visual cognitive styles. *Memory and Cognition, 33*, 710–726.

Kraemer, D. J. M., Rosenberg, L. M., & Thompson-Schill, S. L. (2009). The neural correlates of visual and verbal cognitive styles. *Journal of Neuroscience, 29*, 3792–3798.

Laurillard, D. (2006). *Rethinking university teaching*. Routledge Falmer.

Mason, L. (2018). Multiplicity in the digital era: Processing and learning from multiple sources and modalities of instructional presentations. *Learning and Instruction, 57*, 76–81. https://doi.org/10.1016/j.learninstruc.2018.03.004

Mayer, R. E. (2002). Rote versus meaningful learning. *Theory into Practice, 41*(4), 226–232.

Mayer, R. E. (2009). *Multimedia learning*. Cambridge University Press.

Mayer, R. E. (2014). Incorporating motivation into multimedia learning. *Learning and Instruction, 29*, 171–173. https://doi.org/10.1016/j.learninstruc.2013.04.003.

Mayer, R. E. (2017), Using multimedia for E-learning, *Journal of Computer Assisted Learning, 33*(5), 403–423. https://doi.org/10.1111/jcal.12197.

McEwan, R. C., & Reynolds, S. (2007). *Verbalisers and visualisers: Cognitive styles that are less than equal*. Faculty and Staff Publications.

McKay, E., Garner, B. & Okamoto, T. (2003). 'Management of collaborative web-based experiential learning', paper presented at IASTED International Conference on Web-Based Education, 2003 Rhodes, Greece. *International Association of Science and Technology for Development*, 409–414.

McKay, E. (2000). *Instructional strategies integrating cognitive style construct: A meta-knowledge processing model (Contextual components that facilitate spatial/logical task performance)*, Doctoral Dissertation (Computer Science and Information Systems). Deakin University.

Merrienboer, J. J. G. & Kester, L. (2005). The four-components instructional design model: Multimedia principles in environments for complex learning. In R. E. Mayer (Ed.) *The Cambridge handbook of multimedia learning*. Cambridge University Press.

Messick, S. (1984). The nature of cognitive styles: Problems and promise in educational practise. *Educational Psychologist, 19*, 59–74.

Middleton, C. & Cockrill, P. (2017). Digitally engaged: Experimenting with digital technologies in social history exhibition. *Museums Galleries Australia Magazine, 26*(1), 24–27, Spring-Summer 2017.

Peterson, E. R., Rayner, S. G., & Armstrong, S. J. (2009). Researching the psychology of cognitive style and learning style: Is there really a future? *Learning and Individual Differences, 19*, 518–523.

Price, L. (2004). Individual differences in learning: Cognitive control, cognitive style, and learning style. *Educational Psychology, 24*(5), 681–698.

Rayner, S., & Riding, R. (1997). Towards a categorisation of cognitive styles and learning styles. *Educational Psychology, 17*(1–2), 5–27. https://doi.org/10.1080/0144341970170101.

Rennie, L. J. (2015). John Falk and Lynn Dierking: Building the field of informal/free-choice science education. *Cultural Studies of Science Education*, 11(1), 127–146. https://doi.org/10.1007/S11422-015-9707-3.

Riding, R. (2003). On the assessment of cognitive style: A commentary on Peterson. *Deary and Austin', Personality and Individual Differences, 34*, 893–897.

Riding, R., & Cheema, I. (1991). Cognitive styles an overview and integration. *Educational Psychology, 11*, 193–215.

Riding, R., & Rayner, S. (1998). *Cognitive styles and learning strategies: Understanding style differences in learning and behaviour*. David Fulton Publishers.

Riding, R. J., Grimley, M., Dahraei, H., & Banner, G. (2003). Cognitive style, working memory and learning behaviour and attainment in school subjects. *British Journal of Educational Psychology, 73*(2), 149–169. https://doi.org/10.1348/00070990360626912.

Ryan, J., & McCrae, B. (2005). Assessing pre-service teachers' mathematics subject knowledge. *Mathematics Teacher Education and Development, 7*, 72–89.

Sadler-Smith, E., & Riding, R. (1999). Cognitive style and instructional preferences. *Instructional Science, 27*, 355–371.

Savov, S. A., Antonova, R., & Spassov, K. (2019). Multimedia applications in education. In A. Al-Masri & K. Curran (Eds.), *Smart technologies and innovation for a sustainable future, advances in science, technology & innovation (Ierek interdisciplinary series for sustainable development)* (pp. 262–271). Springer.

Schaller, D. T., Borun, M., Allison-Bunnell, S. & Chambers, M. (2007). One size does not fit all: Learning style, play, and online interactives. *Museums and the Web*, Archives and Museum Informatics.

Schaller, D. T. & Allison-Bunnell. (2007). Practicing what we teach: How learning theory can guide development of online educational activities. In J. Trant & D. Bearman (Eds.), *Museums and the Web*, Archives and Museum Informatics.

Şendurur, E., Ersoy, E., & Çetin, İ. (2018). The design and development of creative instructional materials: The role of domain familiarity for creative solutions. *International Journal of Technology and Design Education, 28*(2), 507–522. https://doi.org/10.1007/s10798-016-9391-y

Seufert, T. (2019). Training for coherence formation when learning from text and picture and the interplay with learners' prior knowledge. *Frontiers in Psychology, 10*(183), 1–11. https://doi.org/10.3389/fpsyg.2019.00193.

Siemens, G., Long, P., Gašević, D. & Conole, G. (2011). *Call for Papers, 1st International Conference Learning Analytics & Knowledge (LAK 2011)*.

Sternberg, R. J. & Grigorenko, E. L. (1997). Are cognitive styles still in style? *American Psychologist, 52*(7), 700–712. https://doi.org/10.1037/0003-066x.52.7.700.

Traxler, J. (2018). Learning with mobiles: The global south. *Research in Comparative and International Education, 13*(1), 152–175. https://doi.org/10.1177/1745499918761509.

Woods, A., & Baker, R. (1985). Item response theory. *Language Testing, 2*, 117–140.

Wright, B. D., & Masters, G. N. (1982). *Rating scale analysis.* Mesa Press.
Wright, B. D. & Stone, M. H. (1999). *Measurement essentials.* Wide Range Incorporation, Delaware.
Wu, M. & Adams, R. J. (2007). Applying the Rasch model to psycho-social measurement: A practical approach. *Educational Measurement Solutions, Melbourne.*

Asmidah Alwi Ph.D., is a senior lecturer in the School of Creative Industry Management and Performing Arts, Universiti Utara Malaysia with 21 years of experience. She received her Ph.D. degree from the RMIT University, Melbourne, Australia. Her area of specialisation is UXD, IXD and multimedia content evaluation. Her research interests span around but not limited to e-Learning and game-based applications with particular interest on human cognitive preferences. Although her work basically involved within the museum and cultural heritage context, Asmidah is keen to venture into any new possibilities. Asmidah is experienced in learning analytics using the Rasch model. Asmidah is an active researcher and has been involved in more than 20 research projects.

ORCID profile https://orcid.org/0000-0002-6040-0255

Exploring a Framework for Constructivist Design with the eLearning Space

Mandi Axmann

Abstract Learning designers and academics require ways to rapidly create eLearning courses. An alternative to time-consuming, content-heavy resources is to design authentic learning experiences that derive from constructivist design. This study aimed to investigate the optimisation of critical learning design elements that informs a constructivist instructional design approach within the Australian eLearning higher education context. A group of subject-matter experts ($n = 53$) from various Australian higher education institutions were purposefully selected to participate in an eDelphi expert panel. An electronic questionnaire was administered to a larger sample ($n = 434$). This study employed exploratory sequential design within a mixed methods approach as narrative and numerical data analysis. The Wilcoxon rank-sum test and Kruskal–Wallis process were applied to validate the perception of respondents identified for the research. The results from the study indicated that learning design elements associated with constructivist concepts such as social learning presence, learning interaction, meta-cognitive load, knowledge-sharing space and knowledge construction may be conducive to the construction of knowledge. The framework presents a modality that can be implemented by instructional designers, academics and online course developers when creating complex eLearning spaces.

eLearning Definition: Purposed for this Chapter

Electronic learning (henceforth referred to as eLearning) can be argued to be a natural extension of distance education. In recent years, mobile technologies (e.g., tablets, mobile phones, smartphones, iPads) have become increasingly popular devices and are now also used to access the web-based modality (Strong et al., 2012). eLearning takes on various modes, such as blended learning, hybrid learning or mixed-mode education. This means that face-to-face (classroom) interaction can be blended with

M. Axmann (✉)
Online Learning Consultant, E-Learning Solutions, Melbourne, Australia

E. McKay (ed.), *Manage Your Own Learning Analytics*, Smart Innovation, Systems and Technologies 261, https://doi.org/10.1007/978-3-030-86316-6_8

asynchronous (not real-time) and synchronous (real-time) methods of computer-mediated communication (Wise et al., 2009). Sangrà et al. (2012) postulated an extended definition of eLearning as an approach to teaching and learning that applies educational models, namely:

> *eLearning ... is based on the use of electronic media and devices as tools for improving access to training, communication and interaction that facilitates the adoption of new ways of understanding and developing learning* (Sangrà et al., 2012).

Despite the exponential growth of eLearning course offerings and this promise of anywhere, anyhow and anytime learning, students continue to report feelings of social disconnectedness (Limniou & Smith, 2010). For students to achieve a sense of connection, they need to establish social learning presence, referring to their ability to project themselves socially and affectively into a learning community.

Constructivist Instructional Design

Constructivist educators view learning as an active process in which learners create new ideas and connections through the reconstruction of experiences (Ammerwerth et al., 2021). Therefore, when engaging in activities online, learners ought to be active and then be given the opportunity to reflect on what they have learnt. Social constructivism was advanced by prominent learning theorists such as Dewey (1902), Piaget (1972), Vygotsky (1978) and Bruner (1990) and is primarily viewed as the active construction of knowledge.

Constructivist instructional design is aimed at constructing eLearning environments in such a way that optimal learning, including generic skills development, may take place (Dick & Carey, 2006; Ellis et al., 2007; Merrill et al., 2008). Collaboration with other learners is therefore essential to create knowledge in the first instance. Individual reflection should be followed up with social networking to check the student's learning progress (Kirschner et al., 2006).

Knowledge Construction and Online Learning Design

The design of online learning activities and social interactions often incorporates technologically supported delivery methods that enables online discussions such as wikis and blogs (Waycott & Gray, 2011). These delivery modes are not always incorporated in a mindful way from the design of the eLearning course (Henderson et al., 2015; Majeski et al., 2015).

Learners ought to construct their own knowledge rather than blindly accepting what is delivered and disseminated by the instructor (Jonassen, 2005). Knowledge construction is facilitated by students completing activities that ensure they took the initiative to learn and interact with other students and the instructor (O'Donnel et al., 2013). It is sometimes difficult to discern which learning design methods promote online social interaction and knowledge construction (Henderson et al., 2015) henceforth the investigation for this study.

Justification for the Study

Standardised lectures that are heavily pre-designed may not always provide opportunities for students to explore real-life problems. The introduction of authentic learning activities is imperative for sense-making and knowledge construction (Waycott & Gray, 2011). Learners are required to be engaged in realistic tasks that provide opportunities for collaborative activities, such as conducting a survey or researching local history (Herrington et al., 2010). An understanding of critical learning design elements to promote social networking for the construction of knowledge and associated learning performance assessment is a key tool for eLearning practitioners (Harris, 2010). This research study contributes to the eLearning domain by presenting an instructional framework that supports knowledge construction within complex eLearning environments.

Emergent models for eLearning show the importance of meta-cognition, collaborative learning, active knowledge-making and recursive feedback as essential components of course design (Cope & Kalantzis, 2013). However, Bradley (2010) observes that when designing a constructivist learning environment, ‘there are no established standards and measurable outcomes, to provide a framework for the instructional designer’ (p. 22). Therefore, despite it being recognised that constructivist learning design is imperative, it is not always clear how these elements may be implemented when more than one form of social interaction (such as a combination of wikis, blogs and discussion forums) is utilised to support learning (Waycott et al., 2013).

Aim of the Study

The general aim of the research study was to investigate learning design elements for an eLearning framework that promote the use of emerging technologies for social networking and co-construction of knowledge (Axmann, 2017). The research aimed to examine the effective use of online learning elements that can contribute towards a framework to optimise learning within the Australian eLearning higher education context.

The main objectives of the research study were as follows:

- Investigate the critical learning design elements for online collaborative learning that are informed by a constructivist instructional design approach.

- Validate the effectiveness of learning design elements for the construction of knowledge against the perceptions of eLearning practitioners in the field of Australian higher education.

Methodology: Exploratory Sequential Design

The study started out by investigating the broad range of literature related to emerging trends within the field of eLearning and instructional design and strategies that support social networking and co-construction of knowledge. The literature focussed on the Australian higher education sector as the context of the study.

The target population of the research study consisted of experienced eLearning practitioners (instructional/educational designers, academic developers, online course coordinators, eLearning advisors) within the Australian higher education institutions, that made use of eLearning, either on-campus or off-campus, as part of their learning design practice. eLearning within the context of this study implies that students may be attending classes on-campus (on-campus), be studying part-time (off-campus) or be studying fully online. A total of 12 ($n = 12$) Australian higher education universities were identified for the study, and 53 representatives from the various institutions were identified to participate in the study. This presented an average of 4–5 respondents per university selected by means of purposeful sampling.

Opinions and feedback from an eDelphi expert panel and subsequent semi-structured interviews were employed to identify and explore critical learning design elements that support social networking and the co-construction of knowledge. This phase established the agreed upon definitions, rationale and review criteria for each of the learning design elements (Table 1).

The focus group discussion was aimed at evaluating, rating and rewriting the proposed instructional learning design elements necessary to facilitate social networking in online environments, including delivery modes of face-to-face, blended and fully online.

The researcher developed an electronic survey for a larger number of the sample population ($n = 434$) in which the identified categories were considered in terms of

Table 1 eDelphi process

eDelphi process	Description of research activity
1. Round one: Opinion poll	The *first questionnaire* was sent to the panel of experts asking for opinions involving experiences and a list of recommendations in terms of the proposed guideline document
2. Round two: Opinion poll and discussion	On the second round, *a copy of the collective list* was sent to each expert and the experts were asked to rate or evaluate each item by some criterion

effectiveness, thus being treated as variables (Cresswell & Plano-Clark, 2011, p. 71). The electronic survey contained questions pertaining to the following categories:

- importance of facilitation tasks;
- online facilitation tasks employed;
- effective advice for online facilitation;
- effective learning activities for online facilitation;
- importance of facilitation tasks in motivating and encouraging students to work collaboratively;
- preferable activities for students in the online classroom, and
- importance of organisational support.

For each relevant learning design element item, a Wilcoxon rank-sum test and Kruskal–Wallis test (Hollander & Wolfe, 1999) were performed to investigate if the perception of respondents were different depending on:

- educational levels (Kruskal–Wallis test;)
- years of experience (Kruskal–Wallis test); and
- location (Wilcoxon rank-sum test).

Limitations of the Study

As the study was contextualised within the Australian higher education, further research would be required to establish transferability to other sectors of education such as primary or secondary education. The locations of the universities extended across different states within Australia, namely Victoria (4 universities), New South Wales (1 university), South Australia (1 university) and Western Australia (1 university). The results and findings of this study cannot be generalised or seen as representative of all Australian educational institutions as Australia is comprised of 37 public universities, two private universities and approximately 150 other providers of higher education.

Overview of the Framework

The eLearning framework for the construction of knowledge presents the description, categories and recommended learning activities for each of the learning design elements. The study identified learning design methods pertaining to social constructivism that received substantial empirical data to demonstrate a significant impact on student engagement.

Learning Design Element 1: Social Learning Presence

Social learning presence fosters that important sense of belonging to a group. Educators may reinforce the sense of belonging and self-esteem by ensuring the engagement of learners in the community (Wenger et al., 2002). Jonassen (1999) was a major contributor to the field of designing constructivist environments and commented on the importance of online interactions that required purposeful design. Facilitation of online tasks encourages social learning presence within academic online practice (Kolb, 2014).

Students that are currently entering the higher education system are expecting social interaction from their online learning environments (Wise et al., 2009). Networked communication takes place through webcam, microphone, text, drawings, telephone or file upload. Social learning presence is enabled by personalised blogs, wikis, websites and podcasts, and thus a more meaningful human presence can be established (Bradley, 2010).

A high percentage of panel members (64.7%) rated the review criteria for social learning presence as critical, and a further 11.8% rated it essential that the online learning activities promote meaningful instructor-student and student–student interactions. More than half of the panel members (70.6%) believed that the web-based course design should allow opportunities for students to interact socially with each other in the online environment.

Respondents were asked to rate how frequent the following learning tasks were employed on a regular basis in the online classroom: (1) small-group online tutorials; (2) group work project; (3) individual written assignment; (4) virtual laboratory activity; (5) field work experiment; (6) online quiz; (7) online final examination; (8) blogging activity; (9) wiki activity; (10) webinars and (11) online portfolio activity. The mean responses ranged from 2.03 to 4.03, and the individual written assignment ($M = 4.03$), online quiz ($M = 3.53$), and group work project ($M = 3.01$) were the three most frequently employed learning tasks on a regular basis in the online classroom. Although the individual written assignment received the highest nomination, it is encouraging to observe that more interactive and problem-based techniques such as online quizzes and group work projects are frequently employed.

Fulltime lecturer (respondent 41) replied that it is more important to experiment with technologies that are useful and user-friendly as opposed to those that are clunky or less user-friendly on an ongoing basis:

> *I also experiment with new technologies so that students can experience the latest technologies available then 'choose' whether to continue to use it when they have completed my course.*

Instructional designer (2) observed that a lack of sense of trust in the environment, whether technological or interpersonal, may adversely affect group dynamics and that reasons need to be created for students to socialise:

> *A lack of meaningful reasons to interact can exist within the course (especially socially e.g. focus on getting qualification, not 'making new friends').*

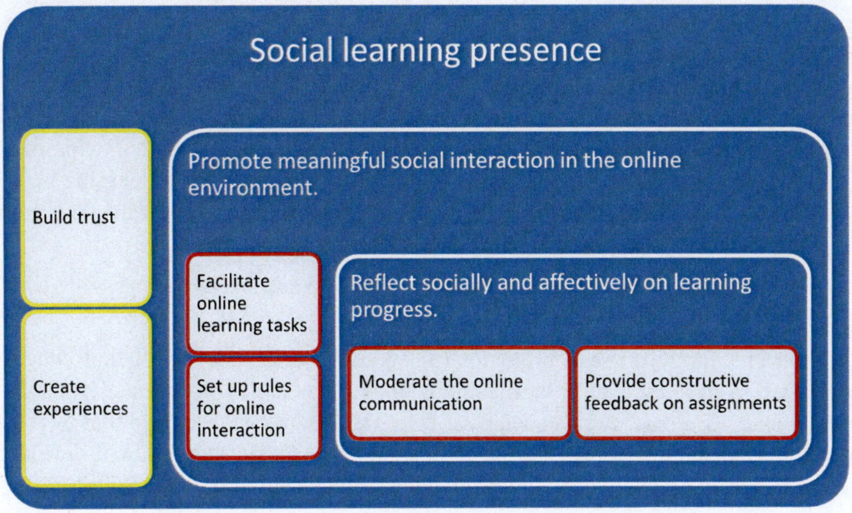

Fig. 1 Summary of social learner presence (Axmann, 2017, p. 199)

Educational designer (12) cautioned that it needs to also be considered that many students are time poor and have existing social networks and that design should acknowledge and incorporate rather than ignore this requirement. Findings from this study for the learning design element 'social learning presence' are summarised by Fig. 1.

Learning Design Element 2: Social Learning Interaction

Social learning presence relies heavily on how the social learning interaction within a course is structured by the attentive inclusion of social media and online communication technologies.

The emergence of disruptive technologies such as mobile learning and Web 2.0 technologies (Cochrane, 2008) also facilitated the move from cognitive pedagogies to social constructivist pedagogies. Engaging the student as an active respondent in education provides a richer environment conducive to student-centred learning (Zeedick, 2010). Educational researcher Gilly Salmon (2000, pp. 25–26) developed a five-step model which outlines the steps required to effectively foster online student engagement through the use of discussion boards.

However, eLearning courses have evolved from discussion-forum led activity, and many courses now also include blogs, wikis and twitter hashtags as additional forms of online communication (Lubega et al., 2014). Salmon's five steps of e-moderating are very sequential and hierarchical and may be sluggish in the current flow of immediacy possible within social media (Majeski et al., 2015).

An important pitfall with constructivism is that too much onus may rest upon the learner to integrate the content (Wheelahan, 2009). With so much knowledge available in the public domain, the learner may not necessarily need to be constructed, but merely assimilate, adopt, critique and/or evaluate the information.

Academic professional developer (4) also remarked that heavy content with little opportunity to interact with others will further adversely affect social learning presence. Senior lecturer in adult education (9) cautioned that social learning was not necessarily seen as part of the academic context, stating:

> *My personal observations are that students like to keep their social and learning contexts separate.*

Online learning course coordinator (3) agreed that it was critical that educational design promoted meaningful interactions between facilitators and students and allowed people to interact socially if they wish. Caution was raised that merely creating opportunities for socialisation does not mean that students will develop online social relationships. As online social presence cannot be assumed or left to chance, the researcher concluded that it is deemed as necessary to include the criteria that the online learning activities provide opportunities for students to reflect socially and affectively on their learning progress.

Respondents were asked to rate the following ten items regarding effective advice for online facilitation to promote engagement in an online classroom: (1) build lecturer-to-student relationship; (2) encourage student-to-student relationships; (3) establish learner support such as 'how to' files and frequently asked questions; (4) create a thread on the discussion forum for informal social interaction; (5) include small-group activities; (6) provide a blog where students can share reflections on their learning experiences; (7) create a wiki page where students can work together on a project; (8) conduct a webinar for real-time student discussion; (9) structure a series of online quizzes based on a case study; and (10) include video presentations of lectured materials. The items related to social learning interaction, namely student-to-student ($M = 4.41$), lecturer-to-student ($M = 4.42$) as well as learner support ($M = 4.26$), rated highest, which further shows how much of a priority these items are to create an engaging and supportive learning space.

Respondents rated including a face-to-face teaching component ($M = 3.97$, SD $= 0.94$) in terms of knowledge construction very highly, indicating that the role of the academic to guide learners is still viewed as extremely important. Face-to-face teaching can also be offered by means of webinars or virtual office (chat) in this context.

The flipped classroom, namely getting students to comment on watching videos, readings and researching articles prior to formal lectures, also received a fair number of comments, such as the following by a sessional lecturer (respondent 45):

> *Flip the classroom and get students to reading, watching videos, researching, reflecting etc. prior to the online sessions so that the online interactions are enriched by the underpinning knowledge.*

A sessional lecturer (respondent 52) stated that it is essential to lead by example if you want to collaborate:

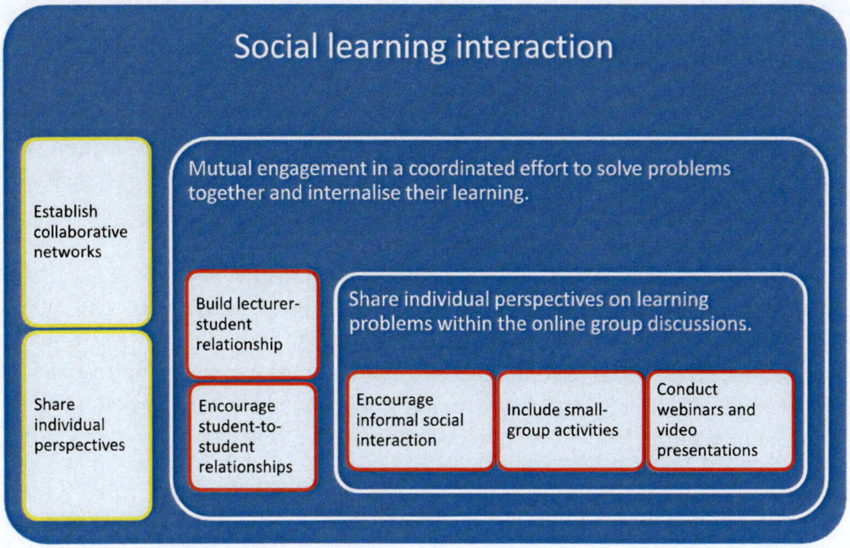

Fig. 2 Summary of social learning interaction (Axmann, 2017, p. 202)

Lead by example. If you want collaboration, start it and praise it.

Respondents also commented further on online collaboration, support and motivation and showing students that you care, which is important for effective social interaction.

The learning design element 'social learning interaction' is visually represented by Fig. 2.

Learning Design Element 3: Knowledge-Sharing Space

Integral to the constructivist process is the development of students' critical thinking skills, problem-solving abilities and teamwork capabilities (Bradley, 2010). Professional discipline knowledge requires the development of generic skills applied in context.

Social learning interaction allows students to demonstrate competence within a peer-to-peer or collaborative setting that would be representative of the workplace environment (Rennie & Morrison, 2013). Caskurlu et al (2021) furthermore identified course design, instructors and peer actions as the main categories that inform the experiences of online learners.

Knowledge-sharing space is viewed as a mostly useful, essential and critical activity. It appeared that panel members rated higher the individual contribution of students (70.6% critical) as opposed to the collaborative activities of interacting, building relationships and a sense of belonging (58.9% critical).

Respondents were asked to rate the following ten items regarding effective activities of online facilitation to encourage students to find solutions to real-life problems in an online classroom: (1) case studies; (2) field work experiments; (3) virtual laboratory task; (4) research activities; (5) virtual simulations; (6) online games; (7) quizzes; (8) virtual tours; (9) written assignments; and (10) examinations. All items had means above 3. Items 1, 2 and 4 had means over 4, indicating respondents viewed these items as moderately to highly effective activities for online facilitation to encourage students to find solutions to real-life problems.

Situated and contextualised learning such as case studies ($M = 4.17$), field work ($M = 4.05$) and research activities ($M = 4.10$) are viewed by respondents as contributing to authentic scenarios that help students find solutions to real-life problems, whereas examinations ($M = 3.25$) were viewed as less important. An online tutor (respondent 47) observed that these types of activities over time are beneficial to both students and tutors in helping them grow and develop ideas and concepts:

Seeing students' ideas and understanding of concepts develop over time, seeing them interact with and support each other, is very rewarding.

Online learning course coordinator (17) remarked that students need to protect their intellectual property and ideas need to be acknowledged as an inhibiting factor to knowledge sharing:

Fear of being ripped off, no clear incentive to share, no protection of intellectual input is a key factor in inhibiting knowledge sharing.

Constructivist environments focus on the organic learning process and cannot be designed in a static, linear fashion (Cope & Kalantzis, 2013). Rather, a space needs to be created for learners where they can engage in reflective practice (Ashman et al., 2012). Course designers ought to debate exactly how much 'new knowledge' is required to be constructed by learners and ensure that assessments are aligned with industry requirements.

This study observed that students in remote communities may still have only dial-up or low-speed internet connections available, and this ought to be considered when designing the student requirements for access to activities and assessments. Situated activities that students can easily participate in and that would not require high-end multimedia design or fast internet connections are evaluation of case studies and reporting on field work experiments. These activities can be structured as small-group activities instead of individual activities to promote knowledge-sharing space.

The learning design element 'knowledge-sharing space' is illustrated by Fig. 3.

Learning Design Element 4: Meta-Cognitive Load

Meta-cognition is a learner's ability to be aware of their cognitive capabilities and use these capabilities to learn (Biggs & Tang, 2011). Jonassen (1999) argues that knowledge construction does not occur in isolation, but is the result of teams of people

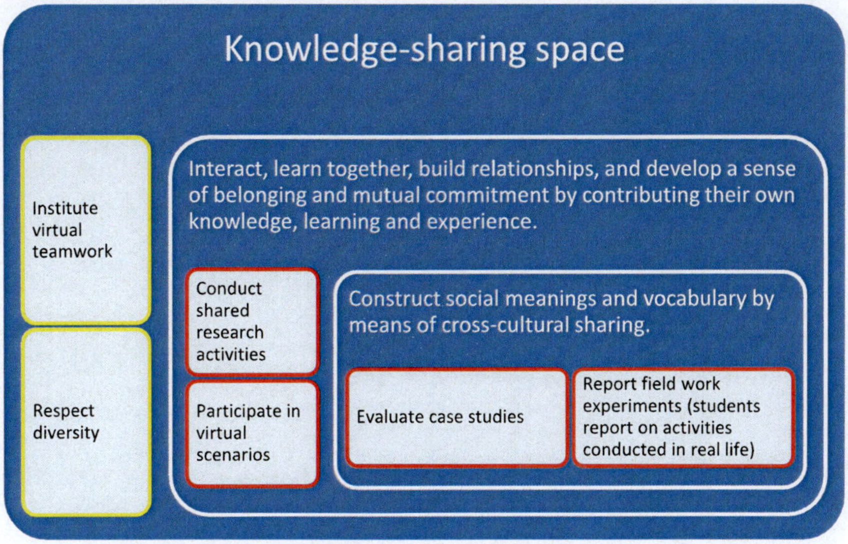

Fig. 3 Summary of knowledge-sharing space (Axmann, 2017, p. 205)

working together to solve a problem, thus the necessity for collaborative online tools. Instructional cognitivist paradigms encourage learners to use meta-cognitive skills to help in the construction of knowledge (Ally, 2007).

Meta-cognitive load activities such as online opportunities for self-monitoring, goal setting, problem-solving and self-reward were valued highly by a substantial percentage of panel members (47.1% essential and 23.5% critical). There is a strong need for active, enquiry based learning that fosters twenty-first century employability skills in graduates (Fraser et al., 2014). Quality manager for distance learning (14) agreed that in the knowledge economy, workers require high level meta-cognitive skills that allow thinking, analysing, problem-solving, research and evaluation.

When students are having trouble understanding the didactic function of different (types of) external representations, this probably reflects meta-cognitive knowledge deficits more than cognitive deficits (Schwonke, 2015, p. 176). Structuring learning activities in such a fashion that meta-cognitive load is managed can help learners gain deeper understanding, acquire knowledge and develop skills quicker (Cope & Kalantzis, 2013). An example of a meta-cognitive tool would be some sort of overview (e.g., a table) of accomplished and open tasks (to facilitate monitoring and planning) (Schwonke, 2015).

Despite the advancements in technology within eLearning, it remains difficult and often frustrating to determine student engagement. A fulltime lecturer (respondent 28) complained that it is more difficult to determine student engagement in an online setting than in a face-to-face environment:

At least we know when on-campus students don't turn up.

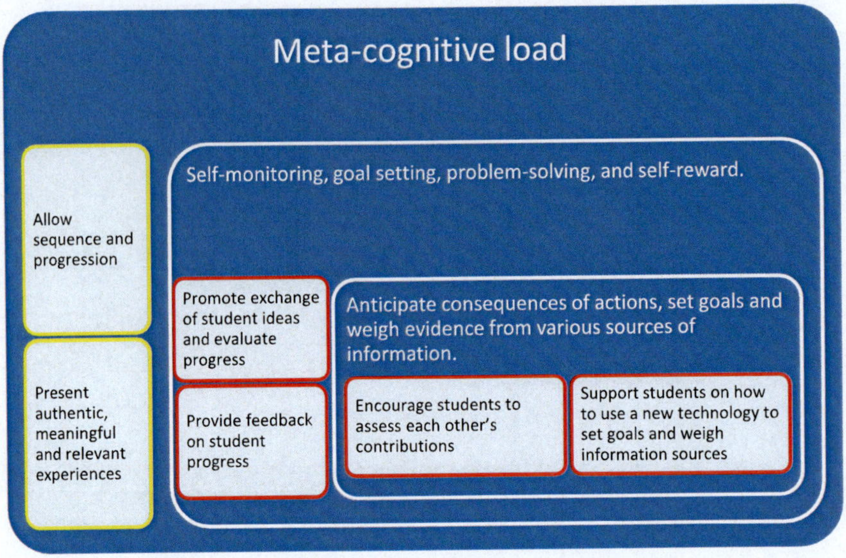

Fig. 4 Summary of meta-cognitive load (Axmann, 2017, p. 207)

The perception remains amongst academics that some disciplines do not lend themselves towards the online environment, as a fulltime lecturer (respondent 98) commented:

> *The discipline I teach into has better outcomes with face-to-face tutorials. Online is harder as the work produced by students (drawings etc.) is not easily accessed on a computer.*

Help files need to be incorporated within the course design to support students on how to use a new technology. Students also need to be supported to set goals and weigh information sources in accordance with their validity and relevance to the learning content.

The learning design element 'meta-cognitive load' is summarised by Fig. 4.

Learning Design Element 5: Knowledge Co-construction

Constructivism anchors the concept of knowledge in the human being (individually and socially) (Von Glaserfeld, 1995), and allowed instructional design to move towards a more human-centred view of information presentation (Ertl, 2010). Knowledge is viewed as living in the human act of knowing; tacit as well as explicit; social as well as individual; and dynamic and cannot be reduced to an object, but is considered a 'human factor' (Wenger et al., 2002).

Knowledge construction includes the experience of meaning-making as the process by which we experience the world and our engagement with it as meaningful (Wenger et al., 2002). Choice and autonomy are important components of meaning-making (Cope & Kalantzis, 2013), and online learning activities and experiences are to be designed in such a way that promotes knowledge construction and transference across various authentic scenarios ranging in complexity (Kolb, 2014).

Most of the panel members agreed that the online learning events need to allow students opportunities to create new knowledge by using appropriate cognitive processes which will improve their ability to exploit existing knowledge (41.1% as useful, 35.2% as essential). Instructional designer (8) suggested that providing space for students to think through difficult concepts was found to be essential to any educational environment. Instructional designer (11) commented that student motivation and the relevance of the learning task also need to be considered.

Findings from this study showed that the expectations for eLearning design are often in stark contrast with organisational policies and restrictions. The concern was raised that if a staff member wants to try a new activity or develop a new assessment item, there are often policy restrictions placed on them in the unit outline (Biggs & Tang, 2011). There is an inherent tension that remains between the learning design, the learning process and the use of the technology (Barton et al., 2009). Students may be offered a choice as to how they would like to present their assignments (such as choosing between video, PowerPoint, or a written assignment). When offering such choice, it is important the grading rubric is specific so that students understand the criteria that they will be graded against.

Most panel members (70.6%) found it essential that the online learning design contains pathways that enable students to identify existing and accessible knowledge and apply this knowledge to solve specific tasks more efficiently. The educator's task therefore is to create a learning environment that structures the learning activities and learner pathway in such a way that this knowledge construction may be achieved (Biggs, 2003).

Further comments by Academic manager for online programmes (26) included that the review criteria were not always easy to easy understand and that too much 'jargon' may have affected the responses. Not all panel members agreed (rating of 17.6%) with review criteria 5.2, namely the online learning design enabled students to identify existing and accessible knowledge. Identification of knowledge was an essential component of the transfer process and necessary to apply the knowledge to solve specific tasks more efficiently and effectively.

These restrictions can sometimes negatively impact on the implementation of novel ways of using the technology for knowledge construction. The learning design element 'knowledge co-construction' is shown by Fig. 5.

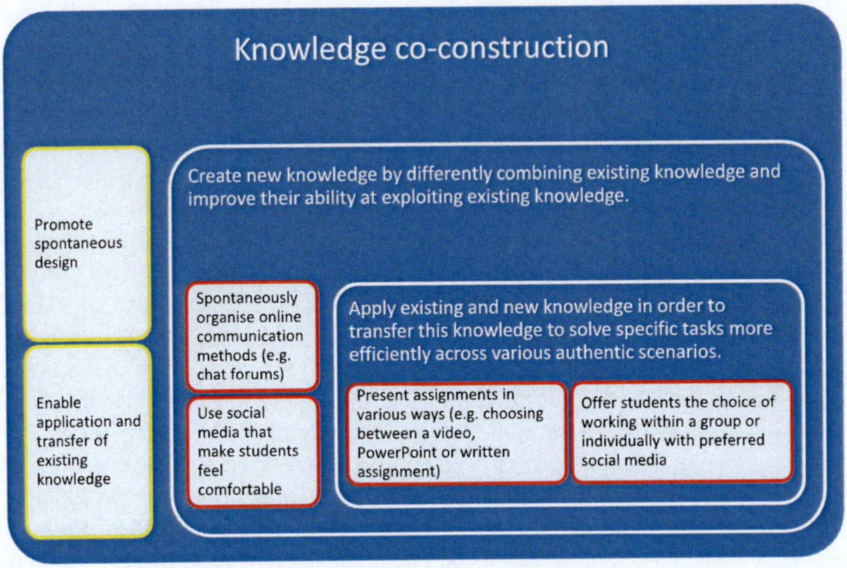

Fig. 5 Summary of knowledge co-construction (Axmann, 2017, p. 209)

Figure 6 indexes the themes that emerged when respondents were asked what they enjoyed most about teaching online.

Multimedia inclusions refer to items such as video, voting functions and quizzes. The technology can facilitate ease of use if the learning environment is set up at the start of the semester. A programme manager (respondent 2) observed that it is easy

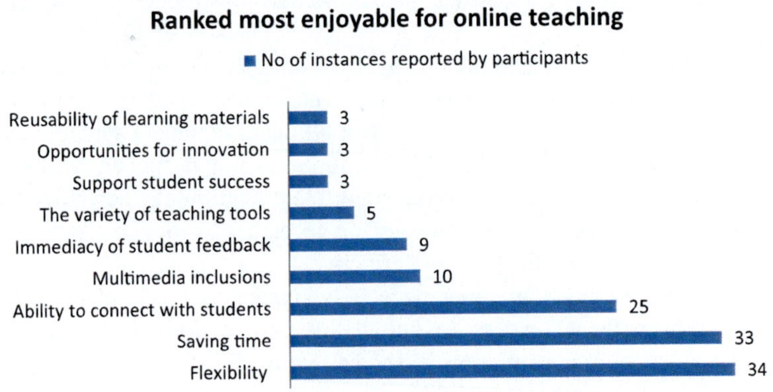

Fig. 6 Ranked most enjoyable for online teaching (Axmann, 2017, p. 189)

to set work up in advance and not to be concerned with students who are unable to attend class due to delays or work and family commitments.

An important aspect of online learning is the ability to interact with students in ways that are not always possible in a face-to-face classroom situation. A fulltime lecturer (respondent 47) commented that it is satisfying to see students grow in confidence.

Respondents rated the flexibility of online teaching very highly. A sessional lecturer (respondent 50) remarked that the flexibility and methods of communication are much more varied when online delivery is included:

Flexibility, time to develop and solve problems when complex questions or situations arise. Number of communication technologies that can be used to communicate with students.

Technology seems to pose challenges for academic staff, but it is not always seen in a negative light. A sessional lecturer (respondent 49) observed that the challenge remains to create effective and enjoyable learning experiences within the online delivery mode. A fulltime lecturer (respondent 47) commented that in his opinion the main advantage of online delivery is that everything is recorded and can be referred to over time.

Everything is recorded and therefore [it is] a very effective way of building resources over time and keeping them organised with minimal effort. [It] also facilitates compliance with various regulatory requirements.

Figure 7 indexes the themes that emerged when respondents were asked what they rated least enjoyable for online teaching.

The online environment can sometimes feel isolating and distant from real-life encounters. A programme manager (respondent 3) observed that he missed the light and informal face-to-face social interaction with students:

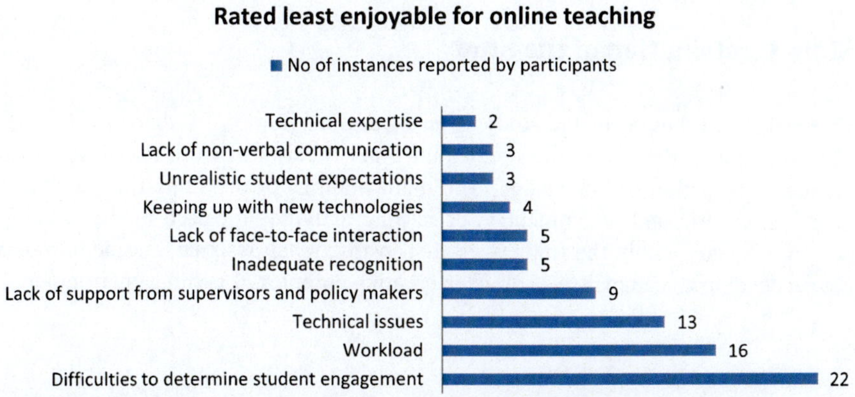

Fig. 7 Rated least enjoyable for online teaching (Axmann, 2017, p. 190)

> I miss the banter with students in the corridors, in the coffee breaks, and some of the ad hoc topics that can come up in a physical classroom.

Organisational support and recognition was a theme that emerged again, and a sessional lecturer (respondent 50) remarked that universities often do not recognise the time and effort it takes to interact with students online, and how demanding student emails and communication may be:

> My experience of online teaching is that it takes more rather than less time compared to face-to-face teaching. The way in which payment is allocated tends to suggest that universities do not expect (and do not want to encourage) all students to participate in online units. When you have one tutor/convenor dealing with 100–200 students, there is no way that tutor/convenor can interact with all of them equally.

Proposed Framework for Construction of Knowledge Within eLearning

The framework postulates that learner presence can be supported by collaborative networks and building trust. Social interaction is promoted by participation focussed on the construction of knowledge. Meta-cognitive load processing is supported by sequence and progression, as well as authentic and purposeful assessments. A sound knowledge-sharing space is encouraged by the establishment of virtual teams that promote respect for diversity. All these elements combined create an optimal space for the transfer of knowledge.

The framework presents a way in which learning design elements can be implemented within an eLearning environment, such as Canvas, Moodle or any related LMS. eLearning practitioners in the field of Australian higher education identified how these learning design elements can be actioned, as described as summarised by Fig. 8.

Main Contribution of the Study

The main contribution of the study is an eLearning framework that instructional designers, course developers, academics and other eLearning practitioners can utilise for the construction of knowledge. The main findings of the study offer a shared meaning, criteria and recommended eLearning activities for each of the learning design elements within the framework that instructional designers, academics and course developers can make use of when creating complex eLearning environments.

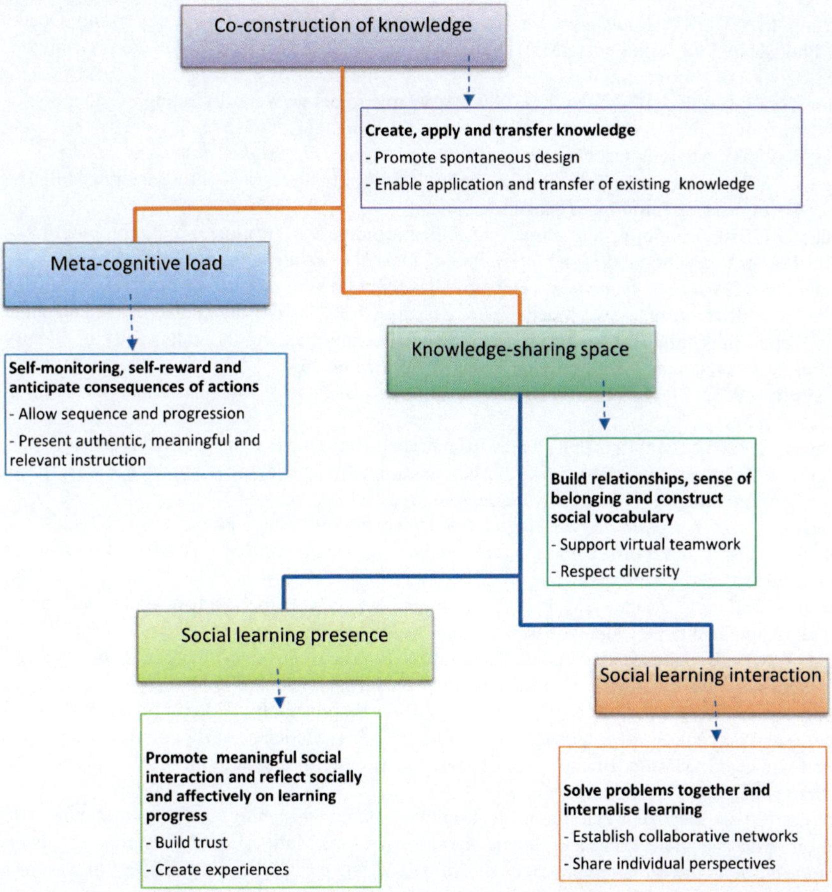

Fig. 8 Framework for the Framework for the construction of knowledge within eLearning (adapted from Axmann, 2017, p. 210)

References

Ashman, H., Dagger, D., Brailsford, T., Goulding, J., O'Sullivan, D., Schmakeit, J. F., & Wade, V. (2012). *Humancomputer interaction and the web.* In J. A. Jacko (Ed.), The human-computer interaction handbook: Fundamentals, evolving technologies and emerging applications (3rd ed.). New York: CRC Press.

Ammenwerth, E., Hacki, W. O., Hoerbst, A., & Felderer, M. (2021). Indicators for cooperative, online-based learning and their role in quality management of online learning. *Research anthology on developing effective online learning courses,* (Vol. 4). Information Resources Management Association.

Ally, M. (2007). Foundations of educational theory for online learning. In T. Anderson & F. Elloumi (Eds.), Theory and practice of online learning (2nd ed.). Athabasca University.

Axmann, M. (2017). *A Framework for Co-construction of Knowledge Within eLearning,* Doctor of Philosophy Full fulfilment, RMIT University. http://researchbank.rmit.edu.au/eserv/rmit:9921864024601341/Axmann.pdf.

Barton, S., Corbitt, B., & Nguyen, L. (2009). Academic social networks affecting the adoption of eLearning in Turkey. *International Journal on e-Learning, 8*(4), 561–575.

Biggs, J. (2003). Aligning teaching for constructing learning. *The Higher Education Academy.*

Biggs, J., & Tang, C. (2011). Teaching for quality learning at university (4 ed.): McGraw Hill: The society for research into higher education.

Bradley, J. (2010). Promoting and supporting authentic online conversations—Which comes first—The tools or instructional design? *International Journal of Pedagogies, 5*(3), 20–31.

Bruner, J. S. (1990). *Acts of meaning.* Harvard University Press.

Caskurlu, S., Richardson, J. C., Maeda, Y. & Kozan, K. (2021). The qualitative evidence behind the factors impacting online learning experiences as informed by the community of inquiry framework: a thematic synthesis. *Computers & Education, 165.*

Cresswell, J. W., & Plano-Clark, V. L. (2011). *Designing and conducting mixed methods research* (2nd ed.). Sage.

Cochrane, T. (2008). Mobile 2.0: The new frontier. Paper presented at the ASCILITE.

Cope, B., & Kalantzis, M. (2013). Towards a new learning: The scholar social knowledge workspace, in theory and practice. *E-Learning and Digital Media, 10*(4), 332–356.

Dewey, J. (1902). *The child and the curriculum.* University of Chicago Press.

Dick, W., & Carey, L. (2006). *The systematic design of instruction* (6th ed.). HarperCollins College Publishers.

Ellis, R. A., Jarkey, N., Mahony, M. J., Peat, M., & Sheely, S. (2007). Managing quality improvement of eLearning in a large, campus-based university. *Quality Assurance in Education, 15*(1), 9.

Ertl, B. (2010). E-Collaborative knowledge construction: *Learning from computer-supported and virtual environments.* New York: Information Science Reference: IGI Global.

Fraser, K., Richardson, J., & Karpathiou, V. (2014). Employability Skills: Differing Requirements and Perceptions of Regulators, Accrediting Bodies, Students, Academics and Employers for Coursework Masters Business (IT) Programs. *The International Journal of Pedagogy and Curriculum, 20*(2), 27–44.

Harris, L. (2010). Electronic classroom, electronic community: designing elearning environments to foster virtual social networks and student learning. In L. H. J.Martin (Ed.), *Information communication technologies for human services education and delivery: Concepts and cases* (pp. 87–104). IGI Global.

Henderson, M., Selwyn, N., & Aston, R. (2015). What works and why? Student perceptions of 'useful' digital technology in university teaching and learning. *Studies in Higher Education,* 1–13. https://doi.org/10.1080/03075079.2015.1007946.

Herrington, J., Reeves, T. C., & Oliver, R. (2010). *A guide to authentic e-Learning.* Routledge.

Hollander, M., & Wolfe, D. (1999). *Nonparametric statistical methods.* John Wiley and Sons Inc.

Jonassen, D. (2005). Tools for representing problems and the knowledge required to solve them. In S. O. Tergan & T. Keller (Eds.), *Knowledge visualization and information visualization: Searching for synergies.* Springer Lecture Notes in Computer Science, Springer.

Jonassen, D. (1999). Designing constructivist learning environments. In C. M. Reigeluth (Ed.), *Instructional design theories and models: A new paradigm of instructional theory* (Vol. 2, pp. 215–239). Routledge Taylor and Francis Publishers.

Kolb, D. (2014). *Experiential leaning: Experience as the source of learning and development.* Pearson Education.

Kirschner, P. A., Sweller, J., & Clark, R. E. (2006). Why minimal guidance during instruction does not work: An analysis of the failure of constructivist, discovery, problem-based, experiential, and inquiry-based teaching. *Educational Psychologist, 41*(2), 75–86.

Limniou, M., & Smith, M. (2010). Teachers' and students' perspectives on teaching and learning through virtual learning environments. *European Journal of Engineering Education, 35*(6), 645-

Lubega, T. J., Mugisha, A. K., & Muyinda, P. B. (2014). Adoption of the SAMR Model to Asses ICT Pedagogical Adoption: A Case of Makerere University. *International Journal of e-Education, e-Business, e-Management and e- Learning, 4*(2), 106–115.

Majeski, R. A., Stover, M., & Ronch, J. (2015). Making asynchronous online learning more learner-oriented: An integrated conceptual model with applications for course design and instruction. *Educational Gerontology, Null-Null.* https://doi.org/10.1080/03601277.2015.1083389

Merrill, M. D., Barclay, M., & Van Schaak, A. (2008). Prescriptive principles for instructional design. In J. M. Spector, M. D. Merrill, J. V. Merriënboer, & M. P. Driscoll (Eds.), *Handbook of research on educational communications and technology* (pp. 173–186). Lawrence Erlbaum Associates.

O'Donnel, E., Mulwa, C., Sharp, M., & Wade, V. (2013). Web mediated education and training environments: a review of personalised interactive eLearning. In E. McKay (Ed.), *ePedagogy in online learning: new developments in web mediated human computer interaction* (pp. 188–208). IGI Global.

Piaget, J. (1972). Intellectual evolution from adolescence to adulthood. *Human Development, 15*(1), 1–12.

Rennie, F., & Morrison, T. (2013). *e-Learning and social networking handbook: Resources for Higher Education* (2nd ed.). Routledge.

Salmon, G. (2002). *E-tivities: The key to active online learning.* Stylus Publishing Inc.

Sangrà, A., Vlachopoulos, D., & Cabrera, N. (2012). Building an inclusive definition of e-learning: An approach to the conceptual framework. *The International Review of Research in Open and Distance Learning, 13*(2), 145–159.

Schwonke, R. (2015). Metacognitive Load - Useful, or Extraneous Concept? Metacognitive and Self-Regulatory Demands in Computer-Based Learning. *Journal of Educational Technology & Society, 18*(4), 172–184.

Strong, R., Irby, T. L., Wynn, J. T., & McClure, M. M. (2012). Investigating students' satisfaction with eLearning courses: The effect of the learning environment and social presence. *Journal of Agricultural Education, 53*(3), 98–110.

Von Glaserfeld, E. (1995). *Radical constructivism: A way of knowing and learning.* Falmer Press.

Vygotsky, L. S. (1978). *Mind in society: The development of higher psychological processes.* Harvard University Press.

Waycott, J., & Gray, K. (2011). *Web 2.0 Authoring tools in higher education: new directions for assessment and academic integrity.* Retrieved from Sydney.

Waycott, J., Gray, K., Clerehan, R., Hamilton, M., Richardson, J., Sheard, J., & Thompson, C. (2010). Implications for academic integrity of using web 2.0 for teaching, learning and assessment in higher education. *International Journal for Educational Integrity, 6*(2), 8–18.

Waycott, J., Sheard, J., Thompson, C., & Clerehan, R. (2013). Making students' work visible on the social web: A blessing or a curse? *Computers & Education, 68,* 86–95. https://doi.org/10.1016/j.compedu.2013.04.026

Wenger, E., McDermott, R., & Snyder, W. M. (2002). *Cultivating communities of practice: A guide to managing knowledge.* Harvard Business School Press.

Wheelahan, L. (2009). The problem with CBT (and why constructivism makes things worse). *Journal of Education & Work, 22*(3), 227–242.

Wise, A. F., Padmanabham, P., & Duffy, T. M. (2009). Connecting online learners with diverse local practices: The design of effective common reference points for conversation. *Distance Education, 30*(3), 317–338.

Zeedick, D. (2010). *The modified delphi method to analyze the application of instructional design theory to online graduate education.* (Ed.D. 3397676), The George Washington University, United States District of Columbia.

Mandi Axmann completed her Ph.D. at RMIT University, Australia during 2017 with the focus of constructivist design methodologies for multimodal learning spaces. Her work experience extends across Higher Education, Tertiary and Further Education (TAFE) and corporate educational sectors. She currently works as a consultant for best practice in educational design and development. She provides services for educational resource development, instructional design, and professional coaching.

ORCID profile https://orcid.org/0000-0002-4700-6392

Industry 4.0: Learning Analytics Using Artificial Intelligence and Advanced Industry Applications

Mahmoud Numan Bakkar and Mandi Axmann

Abstract Learning analytics is a crucial factor in measuring the effectiveness of any educational programme. Industry 4.0, mainly artificial intelligence (AI) applications are integrated into the digital pedagogical techniques and the learning analytics that further enhance and develop the online instructional materials. Therefore, it is crucial to explain the usage of those technologies used for the learning analytics associated with the educational system to show the different use of AI. This book chapter explains: the meaning of the learning analytics in Industry 4.0 and education and its affiliated areas; AI and its related areas with examples from state-of-the-art corporate applications and solutions based on AI technologies integrated into the learning analytics.

Introduction

The fourth industrial revolution (IR4 or Industry 4.0) surpasses the previous industrial revolutions in its scale, scope and complexity. With the advancements in technology, people can now produce new information and generate knowledge, and subsequently automation enhances the intelligence of machines. Artificial intelligence (AI) gives rise to big data analysis and data mining, machine learning algorithms and validation models, which in turn impacts on the way we live, work and study.

This book chapter provides an overview of learning analytics and its different applications derived from AI research, and how these advancements are making an impact on the educational arena. The potential usage of artificial intelligence in learning analytics is discussed. Finally, the chapter lists different examples and case studies of companies providing up-to-date solutions and provides a SWOT analysis for introducing learning analytics within Industry 4.0.

M. N. Bakkar (✉)
Institute of Applied Technology (IAT), Abu Dhabi Vocational Education and Training Institute, Abu Dhabi, UAE
e-mail: Mahmoud.Bakkar@sts.adveti.ac.ae

M. Axmann
E-Learning Solutions, Melbourne, Australia

E. McKay (ed.), *Manage Your Own Learning Analytics*, Smart Innovation, Systems and Technologies 261, https://doi.org/10.1007/978-3-030-86316-6_9

Industry 4.0

The concept of Industry 4.0 was conceived in Germany in 2011 to solve the issue of using the high technology (cyber-physical systems, Internet of Services (IoS), Internet of Things (IoT) strategies in the German economic policy (Roblek et al., 2016). Machines will be able to stream the data in a real-time manner between each other's by utilising the intelligent systems and the wireless sensors, and the data will be analysed for further added value and personalised services as product customisation by the customers or organisations (Roblek et al., 2016).

The cyber-physical systems revolution coined as the fourth industrial revolution (4IR or Industry 4.0) reaches back to the waves of the industrial revolutions, spanning from water to steam power and electric power to electronics. The transition from hand to machine production brings about a revolution in how technology assist with everyday tasks, travel, industry and business as highlighted in Fig. 1.

Industry 4.0 looks to integrate technologies from different sectors (IT, health care, construction and energy, etc.) (Roblek et al., 2016). Self-smart factories will use optimisation and autonomous systems; intelligent cities will be built on smart cities (economy, mobility, people, governance and living) for more digital sustainability and further efficiency in resource usage (Roblek et al., 2016). IoT is about connectivity with all smart devices.

The digitisation process further enabled the use of wireless technology and the mobile applications, Internet systems and platforms, towards the broad usage of cloud computing, intelligent systems, automation systems protected by the secure cyber technology, all together enabled the evolution towards Industry 4.0. The world economic forum highlighted that the fourth industrial revolution's progress started

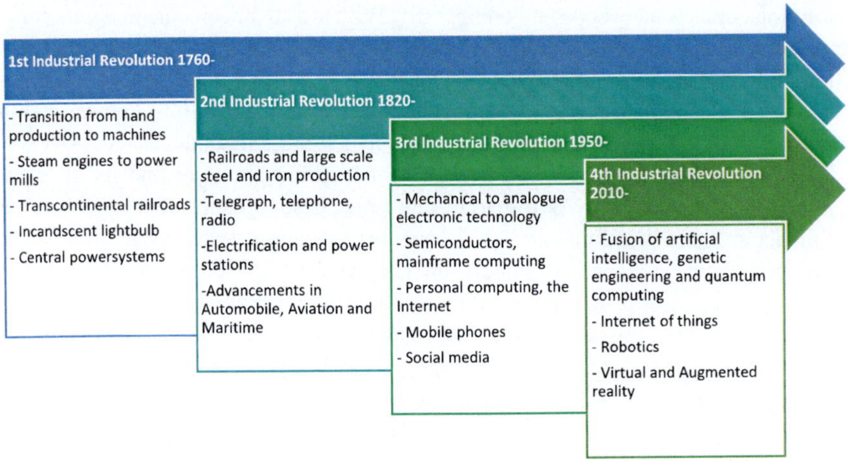

Fig. 1 Overview of the industrial revolutions

since the twenty-first century, and it is radically changing its shape by different aspects, such as mobile internet, stronger sensors, artificial intelligence, and engine learning (Schwab, 2016).

Artificial Intelligence

The advent of artificial intelligence (AI), machine learning (ML), robotics, Internet of Things (IoT), nanotechnology, biotechnology, energy storage and quantum computing are blurring traditional manufacturing boundaries. Autonomous vehicles and self-driving cars, 3D printing, virtual and augmented reality, wearables, additive manufacturing gives rise to new business models, for example Uber and Tesla.

Artificial intelligence involves the activities needed to make the machines act intelligently within their environment (Stone et al., 2016). Computers have been completing ever more complex tasks more speedily, and artificial intelligence (AI) represents technological applications now becoming competent in aspects that previously only thought human brains could undertake, such as playing chess or composing (Christianini et al., 2017). In 2015, the DeepMind's AlphaGo trained with reinforcement learning defeated the human world champion in the ancient game of Go and NicelBot made a song wrote by a bot during 2020.

There are two kinds of AI, namely strong machines that think creatively and deploy deep learning using neural networks and weak AI meaning machines that take on the automation of complex but routine tasks (Salmon, 2019). New devices promise to be less-intrusive and offer-intuitive ways to amplify human intelligence (Salmon, 2019). Artificial intelligence is successful in big data and statistics and making complex calculations using huge quantities of information.

By the mid-2000s, researched experienced a paradigm shift and instead of trying to mimic logic, the algorithms were aimed at calculating statistics, for example make a calculation every time you drag an email to your spam filter. Combining this information, the AI algorithm can now make an educated guess about new emails. Suddenly, machines could engage in predictive behaviour such as completing sentences, predict a next click or recommend a product (Christianini et al, 2017). The predictive ability of machine learning becomes important for learning analytics when looking at student success and learner satisfaction.

Attention was drawn to the fuel of intelligent machines, namely data. A variety of ways to elicit and collect data, namely asking a friend to accept a cookie, tag a friend in an image, rate a product, play a location-based game contributed to developing global data infrastructure. Every time the Internet is accessed to read the news, play a game, check an email or social media feed, and there is interaction with this infrastructure. The infrastructure is not only a physical one of computers and wires, but also of software, including social networks and blogging sites (Christianini et al, 2017).

Table 1 Research on AI

Company name	AI research website
Microsoft	https://www.microsoft.com/en-us/ai/ai-school
Apple	https://machinelearning.apple.com/
Google	https://ai.google/education/
IBM	https://www.ibm.com/training/ailearning
Samsung	https://research.samsung.com/artificial-intelligence
Sony	https://ai.sony/index.html
Facebook	https://ai.facebook.com/

Realising the need for AI in education, many giant companies such as Google, Microsoft, IBM, Apple, Samsung launched a research section focusing on AI, as shown in the Table 1:

Microsoft company set its AI strategy by protecting the ethics of its AI solutions and its deployments based on the below AI principles:

1. **Fairness**

   Having the AI solution acting reasonably with all people in society requires diversity amongst the people developing the AI solution (Responsible AI Principles from Microsoft, 2021) and the measurement highlighted in the rest of this book chapters.

2. **Reliability and safety**

   The designed AI solution should be consistent with the fundamental ideas, values, principles and is used everywhere in a consistent approach; also, the system should be safe and not harm human life (Responsible AI Principles from Microsoft, 2021).

3. **Privacy and Security**

   Since the data drive the AI, the resulting measurement should be secure from any external threat or alteration to increase the reliance on AI solutions. (Responsible AI Principles from Microsoft, 2021).

4. **Inclusiveness**

   To ensure everyone gets the benefits of using AI solutions, all communities and people should be engaged regardless of age, gender, ethnicity, race and mental health; this dogma is the basic principle for a sustainable AI solution. (Responsible AI Principles from Microsoft, 2021).

5. **Transparency**

   The AI solution interpretation of the data measured to take the decision should be revealed how is the decision taken and how the interpretation is made; in addition, the AI solution behaviour should be monitored by the people and not only depending on the machine. (Responsible AI Principles from Microsoft, 2021).

6. **Accountability**

   The AI solution deployment and usage should be upon the people who design it or use it; there should be a human authority that interferes with clearing the limitations and the weakness of the AI solution, so the humans are the ultimate final decision on the AI solution decision. (Responsible AI Principles from Microsoft, 2021).

The principles mentioned above lead to the need for collaborative cognitive research between the education industry, AI solution providers, and psychometric and cognition measurement researchers. However, AI acts as one of the primary Industry 4.0 technologies used in learning analytics. The research into knowing how to use AI in cognition measurement is promising. Different AI areas anticipate the adoption of AI in the learning analytics domain.

Moreover, to integrate AI technologies and the continuous development of applied research in educational institutions, they need to adapt their pedagogical and corporate business models to implement the above principles. Such as openness and having the data reach out to any researcher and users to attract collaboration and excellence, and further scaled modelling.

Below are the different areas in AI promising more integration with the educational cognition measurement and the learning analytics:

1. **Machine Learning**

   Defined as a branch of the AI focusing on data analytics for automating the model building, by figuring out the patterns and deciding with extremely low human interventions, the goal is to understand the data that fits the theoretical model (SAS Institute Inc, 2021). Machine learning will assist education providers in their data analytics by creating a prediction model of the under risk or low-performance students or categorising the student's achievement or weaknesses as their study needs. Machine learning algorithms such as clustering, classification and association expect to be used in data analytics.

2. **Deep Learning**

   Deep learning works on a more significant amount of data using more complex algorithms and more powerful computing to learn multiple levels of abstractions and representations (SAS Institute Inc, 2021). Deep learning algorithms can be further utilised in studying the students' behaviours and assist in the analysis of big educational data.

3. **Robotics**

   Robots are agents equipped with sensors to perceive the environment to perform tasks by perceiving and manipulating the surrounding world—three types of robots: manipulators, mobile robots, unmanned ground vehicles. (Russell & Norvig, 2021). Robots can help the students in tutoring and personalised learning needs. In addition to personal tutoring assistant which assisting the educational institution reduces the workload on the teachers. Such as the use of chatbots currently used by many organisations.

4. **Computer Vision**

   Computer visions enable computers to gain helpful information from digital images and videos to take actions based on the perceived information (IBM, 2021a, 2021b). In addition, systems such as: augmented reality; image processing; face detections; and object detections, can assist in learning analytics, such as proctoring the assessments online.

5. **Natural Language Processing (NLP)**

   Natural language processing focuses on creating computer systems that can understand the human language and the people speech and text the same as the human (IBM, 2021a, 2021b). Its highly needed solutions, especially with special needs students, help them read or speak and learn the languages quickly.

Education 4.0

Billions of people connected by mobile devices with unprecedented processing power, storage capacity and access to knowledge may become a critical domain for knowledge creation and learning (Salmon, 2019). The hive mind or collective consciousness refers to 'a shared way of thinking amongst human beings in the pursuit of knowledge' … (Zukerfield, 2017) and offers new pathways for educational collaboration amongst both students, educators and researchers.

Education 4.0 is not merely the introduction of a new technology but brings about a quiet revolution in how the educational content is presented, structured and assessed (Salmon, 2019). In a similar way that the advancement of the World Wide Web and online learning required new approaches, the progression of artificial intelligence and machine learning is set to revolutionise the educational landscape.

Traditionally, the introduction of new technologies in education is met with a mixture of both excitement and trepidation. Excitement for trying out something new and innovative and a possible reduction in heavy administrative workload. There is also apprehension towards the learning curve and challenge to the status quo that the new technology represents. The growing interest in learning analytics has increased through extensive digitisation in all aspects of our lives.

Learning Analytics

Learning analytics may be defined as... 'the measurement, collection, analysis and reporting of data about learners and their contexts, for purposes of understanding and optimising learning and the environments in which it occurs' (Gašević et al., 2015). Instead of using the typical learning and teaching classroom pedagogies, such as face-to-face teaching and traditional teaching methods such as the chalkboard, the new learning methodologies emphasise the latest digital technology such as: tablets; mobile phones; online conferences; and video call tools, such as Zoom and Microsoft teams.

AI is the central area of Industry 4.0 has been utilised in learning analytics for creating a personalised learning experience for the students by creating personalised course plans by measuring the student's cognition and understanding their ability by identifying their weaknesses and the knowledge skills gabs towards the applications of the knowledge in using the learned skills. Today, student's use of digital content is exponentially increasing. This means, different formats such as graphics, videos, presentations, shared files, online content and live-stream content, all of those mentioned organised for better access using an AI interactive interfaces, which are designed and customised based on the student needs.

AI features such as: machine learning; deep learning; computer vision; robotics; and natural language processing can be all utilised to enhance the quality of online education. Further improvements are possible to the quality of delivery of the intelligent course contents, including: virtual learning environments; personalised learning; innovative content; learning tasks; automation; virtual assistants; facial recognition; and secure online assessments. In addition to that, they are using effective feedback tutoring systems and interactive AI tools to increase student's engagement based on learning theory and cognitive computing.

AI solutions came to service students in K-12, higher educations, personal training. Different companies work in this field, such as: Carnegie learning who provide learning solutions for K-12 mathematics; literacy; world languages; and professional learning. Cognii—AI for education offers K-12 and higher education students such as one-to-one AI-powered tutoring for higher-order thinking and critical thinking; direct feedback and scoring; and formative assessment to identify the students' knowledge commentary. Also, by focusing on persistence knowledge by engaging the learning experience and uses assessment tools that reduce the workload on the teachers. In addition to the analytics of the student's interactions and the personalisation of the student's experience.

Another company name is Nuance, which uses speech recognition to enhance the learning experience. A company named Century Tech, worked on designing AI-driven personalised learning. Kidaptive uses AI to create an adaptive learning platform that supports the educational companies to analyse the data collected to increase the student's engagement and improve the outcome of the learning experience using the learner behaviour metrics. Blippar uses augmented reality, which combines the digital and the real world in training and visualisation by showing interactive materials to improve memory recall. Another company name Quizlet provides flashcards, diagrams and study guides.

However, using the latest technology increased the demands for evaluating the effectiveness of using such tools. Because of this, learning analytics plays a crucial aspect in solving this problem. For some, learning analytics is about knowing how to collect large amounts of data to support future enhancement and improvement based on educational data mining and technology-enhanced learning (Friedl et al., 2020).

Learning analytics relates to Industry 4.0 technologies like AI, machine learning and big data analysis in the education industry to automate further the effective reporting of the learning and teaching methodology, such as online engagement and cognition measurement.

Learning analytics concerns gathering the user performance interaction data that is logged and extracted. These metrics are used to evaluate the learning pedagogy effectiveness to identify student weaknesses. These data provide student feedback teachers concerning their virtual learning environment delivery, affording further improvement (Friedl et al., 2020). Whilst predictive analytics gives an indication of what may happen, prescriptive analytics shows the way to make it happen. Predictive analytics can be defined as the use of statistical and modelling techniques to determine what may happen in future (El Morr & Ali-Hassan, 2019).

Moreover, the learning analytics factors for the student's study success (Yau & Ifenthaler, 2020) mentioned that students' gender and backgrounds, such as their ethnicity, cognition ability and their prior knowledge, study performance and their attributes such as their motivations and the psychosocial contextual, factors related to the course such as active learning and attentions and environmental factors such as supportive academics. In closing discussion on Industry 4.0 digital technologies and their AI-related areas, it is projected that future psychometric measurement software packages will use real-time data (AI Data-driven) to bestow better educational design making decisions for students and teachers using real-time learning analytics.

SWOT Analysis of Industry 4.0 and the Impact on Learning Analytics

The World Economic Forum predefined ten employability skills required by 2020 with the main theme 'Employment, Skills and Workforce Strategy for the Fourth Industrial Revolution'. The report stated ten skills that will be required in the job market by employers by the year 2020. These skills included complex problem solving, analytical thinking, innovation, critical thinking, creativity, emotional intelligence, system analysis, evaluation, leadership and social influence, judgement and decision-making (Paucenau et al., 2020).

However, it is also expected that the fourth industrial revolution will have a massive impact on careers and the job market by changing the basis of work, employment and business (Pauceanu et al, 2020). Moreover, it is also forecasted that due to significant development in artificial intelligence, one-third of today's jobs will disappear by 2025 (Brougham & Haar, 2017). Thus, educational institutions are required to place a much higher premium on student's employability skills.

How do we connect the dots between Industry 4.0, preparing our students for the new world of work and learning analytics? Can learning analytics aid us to better understand the needs of our students and improve curriculum to prepare learners for increasing demanding work environments?

A wide body of literature suggests that predictive learning analytics (PLA) can assist educators to identify students that may need extra support and suggest actionable interventions and proactive support, for example predict which student may be at risk for not submitting their next assignment (Herodotou et al, 2019). Researchers from various fields should be encouraged to work together to address the complex issues in education.

To make sense of the potential contribution of learning analytics and advanced industry applications for Industry 4.0, the authors compiled a SWOT (strengths, weaknesses, opportunities and threats) analysis of the current challenges as a part of the literature review as illustrated by Table 2.

In summary, the strengths of deploying learning analytics within coursework and curriculum are apparent and may contribute to the analysis of work-related skills, identify gaps in academic curriculum and industry expectations, improve the teaching and learning process and advance deep learning with personalised learning experiences. There is also opportunity to create personalised learning experiences by extrapolating factors for student success, motivation and support. Predictive learning analytics is an innovative student's support approach in online pedagogy that can empower online teachers in effectively monitoring and intervening with their students (Heredotou et al., 2019).

Table 2 Comparing the strengths, weaknesses, opportunities and threats for learning analytics and Industry 4.0

Learning Analytics and Industry 4.0	
Strenghts - Analyze work-related skills - Identify gap in academic curriculum and industry expectations - Improve teaching-learning processes - Include industry and workplace dynamics in assessments - Advance deep learning - Create personalised learning experience - Extrapolate factors for student success, motivations and support - Provide meaningful predictive and prescriptive analysis of learner data - Identify students at risk	**Weaknesses** - Additional academic professional development to make use of new technology - Increase expenses of advanced technologies - Workload related to creating personalised learning environments, such as content and course creation - Continuous transformations in industry roles and expectations - Reductionist viewpoint of individual student components - Actions should be taken to motivate and engage online educators to make use of the technologies
Opportunities - Contribute to development of the global economy - Advance employability skills in cooperation with industry stakeholders - Graduates need experience, solve complex problems, correct attitude and aptitude for changing job categories - Design thinking, creativity, ICT use, problem solving, data and digital mangement and user experience (UX) becoming the norm in marketplaces -Increase ability of educational institutions to prepare students for employment - Multi-disciplinary focus to address challenges	**Threats** - Graduate skills shift from technical to soft skills such as social interaction, confidence and self-reflection - Employers hesitant to invest in training and development - Graduates not equipped for modern workplace - More competitive labour market - Graduates ability to be comfortable with disruption, change and uncertainty - Change in educational culture and curriculum - Student privacy and ethical collection

However, the implementation of learning analytic tools is not without expense, and additional professional development and increased workload and the continuous changes in industry expectations are also a consideration. There is also a danger that a reductionist view of the student could fail to see all the student components contributing to student success. Actions should be taken to encourage educators to engage with learning analytics technology and apply the results to their teaching practices (Balica, 2018; Islam et al., 2021; Mubarak et al., 2021; Prinsloo & Slade, 2017; Ranjeeth et al., 2020; Renties et al., 2020).

The opportunities of contributing to the advancement of a global economy, advance employability skills, enable students for complex problem solving and digital management are worthy causes for educational institutions. Design thinking, creativity, ICT use, problem solving, data and digital management and user experience (UX) are becoming the norm in marketplaces, and inclusion of these practices in the curriculum may increase ability of educational institutions to prepare students for employment.

These factors weigh up against the threats of employers hesitant to invest in training and development. Graduate skills shift from technical to soft skills such as social interaction, confidence and self-reflection, and graduates may not be equipped for the modern workplace if changes to the educational curriculum are not implemented in alignment with industry expectations. Education also needs to support graduates' ability to be comfortable with disruption, change and uncertainty. Last and not least, student privacy and ethical collection of learning analytics data always ought to be carefully considered when implementing any learning analytics system (Islam et al., 2021; Mubarak et al., 2021; Prinsloo & Slade, 2017; Ranjeeth et al., 2020).

Conclusion

This chapter discussed the advancements of the fourth industrial revolution (4IR or Industry 4.0) and the impact on education and learning analytics. There is a need for collaborative research between education, industry, AI solution providers to come up with solutions for improved learning analytics that feeds into the learner experience and employability.

The student use of digital content is increasing exponentially, and AI features such as machine learning, deep learning, computer vision, robotics and natural language processing can be all utilised to enhance the quality of online education. The implementation of learning analytics is not without expense and consequence to the change in educational culture and curriculum and can be weighed up against the ability of educational institutions to prepare students for future world of work.

References

Balica, R. (2018). Big data learning analytics and algorithmic decision-making in digital education governance. *Analysis and Metaphysics., 17*, 128–133.

Brougham, D., & Haar, J. (2017). Smart technology, artificial intelligence, robotics, and algorithms (STARA): Employees' perceptions of our future workplace. *Journal of Management & Organization, 24*(2), 239–257. https://doi.org/10.1017/jmo.2016.55

Christianini, N. Norvick, P. & Graham-Cumming, J. (2017). The challenge of creating intelligent machines. In D. Heaven (Ed.), *Machines that think: everything you need to know about the coming of ate of artificial intelligence.* John Murray Learning.

El Morr, C., Ali-Hassan, H. (2019). Descriptive, predictive, and prescriptive analytics. In *Analytics in Healthcare. SpringerBriefs in health care management and economics.* Springer. https://doi.org/10.1007/978-3-030-04506-7_3.

Friedl, M., Ebner, M., & Ebner, M. (2020). Mobile learning applications for android and iOS for German language acquisition based on learning analytics measurements. *International Journal Learning Anal Artificial Intelligence Education*

Gašević, D., Dawson, S., & Siemens, G. (2015). Let's not forget: Learning analytics are about learning. *TechTrends, 59*(1). https://doi.org/10.1007/s11528-014-0822-x.

Herodotou, C., Hlosta, M., Boroowa, A., Rienties, B., Zdenek, Z., & Mangafa, C. (2019). Empowering online teachers through predictive learning analytics. *British Journal of Educational Technology., 50*(6), 3064–3079.

IBM. (2021a). *What is computer vision*? [WWW Document]. https://www.ibm.com/topics/computer-vision (Accessed February 7, 2021).

IBM. (2021b). *What is natural language processing*? URL https://www.ibm.com/cloud/learn/natural-language-processing (Accessed February 7, 2021).

Islam, S., Mouratidis, H., & Mahmud, H. (2021) An automated tool to support an intelligence learner management system using learning analytics and machine learning. In I. Maglogiannis, J. Macintyre, L. Iliadis (Eds.), *Artificial intelligence applications and innovations. AIAI 2021. IFIP advances in information and communication technology*, p. 627. Springer. https://doi.org/10.1007/978-3-030-79150-6_39.

Kanuru, S. L., & Priyaadharshini, M. (2020). Lifelong learning in higher education using learning analytics. *Procedia Computer Science, 172*, 848–852. https://doi.org/10.1016/j.procs.2020.05.122

Microsoft AI principles [WWW Document]. (2021). *Responsible AI principles from microsoft.* https://www.microsoft.com/en-us/ai/responsible-ai?activetab=pivot1%3aprimaryr6 (Accessed January 7, 21).

Mubarak, A. A., Cao, H., & Ahmed, S. A. (2021). Predictive learning analytics using deep learning model in MOOCs' courses videos. *Education and Information Technologies, 26*, 371–392. https://doi.org/10.1007/s10639-020-10273-6

Pauceanu, A. M., Rabie, N., & Moustafa, A. (2020). Employability in the fourth industrial revolution. *Economics and Sociology, 13*(3), 269–283. https://doi.org/10.14254/2071-789X.2020/13-3/17

Prinsloo, P. & Slade, S. (2017). An elephant in the learning analytics room: the obligation to act. In *Proceedings of the seventh international learning analytics & knowledge conference,* pp. 46–55.

Ranjeeth, S., Latchoumi, T. P., & Victer, P. (2020). A survey on predictive models of learning analytics. *Procedia Computer Science, 167*, 37–46. https://doi.org/10.1016/j.procs.2020.03.180

Renties, B., Simonsen, H. K., & Herodotou, C. (2020). Defining the boundaries between artificial intelligence in Education, Computer-supported collaborative learning, Educational data mining and learning analytics: A need for coherence. *Frontiers in Education, 5*, 128. https://doi.org/10.3389/feduc.2020.00128

Roblek, V., Meško, M. & Krapež, A. (2016). A complex view of Industry 4.0. *Sage Open*, 6, 2158244016653987.

Russell, S. & Norvig, P. (2021). Artificial Intelligence: A modern approach, *Global Edition 4th. Foundations 19*, 23.

Salmon, G. (2019). May the Fourth Be with You: Creating Education 4.0. *Journal of Learning for Development,* 6(1), 95–115.

SAS Institute Inc. (2021). Machine Learning: What it is and why it matters | SAS [WWW Document]. Machine Learning: What it is and why it matters|SAS. https://www.sas.com/en_us/insights/analytics/machine-learning.html#:~:text=Machine%20learning%20is%20a%20method,decisions%20with%20minimal%20human%20intervention.

Schwab, K. (2016). *The fourth industrial revolution. Journal of Career Assessment.*

Stone, P., Brooks, R., Brynjolfsson, E., Calo, R., Etzioni, O., Hager, G., Hirschberg, J., Kalyanakrishnan, S., Kamar, E. & Kraus, S. (2016). Artificial intelligence and life in 2030: the one hundred year study on artificial intelligence. *Report of the 2015–2016 Study Panel,* Stanford University, Stanford, CA, September 2016. Doc: http://ai100.stanford.edu/2016-report.

Teng, W., Ma, C., Pahlevansharif, S., & Turner, J. J. (2019). Graduate readiness for the employment market of the 4th industrial revolution: The development of soft employability skills. *Education and Training, 61*(5), 509–604.

Yau, J.Y.-K., & Ifenthaler, D. (2020). Reflections on different learning analytics indicators for supporting study success. *International Journal of Learning Analytics and Artificial Intelligence for Education: IJAI, 2,* 4–23.

Zukerfeld, M. (2017). How to know knowledge? Introducing cognitive materialism. *knowledge in the age of digital capitalism: An introduction to cognitive materialism.* University of Westminster Press.

Mahmoud Bakkar has a Ph.D. in Information Technology/Computer Engineering and Business Information systems from RMIT University—Australia. He has a Master degree in Information Technology Management, UOW—Australia, Bachelor of Computer Engineering, JUST- Jordan. He has extensive experience in teaching and training in the UAE and Australia. He taught courses for both undergraduate and postgraduate levels in Australia and UAE. Also, he has a vast knowledge of business development, management, and leadership. Dr. Bakkar's current research interests include: Artificial Intelligence Systems; Industry 4 applications; Industry 5 applications; Smart City Systems and Innovations; Cloud Computing; Cyber Security; IT in Education; Assessment Measurement; Data Mining; Big Data and Business Intelligence Analytics; Cognitive Computing and their application in Business; IT Innovation and Management; Instructional Design; Online Training and Educations; E-Commerce; and Online Business.

ORCID profile https://orcid.org/0000-0003-4637-4035

Mandi Axmann Ph.D., Online Learning Consultant, e-Learning Solutions linkedin.com/in/dr-mandi-axmann-55535112.

ORCID profile https://orcid.org/0000-0002-4700-6392

Machine Learning for Learning Analytics for Meta-cognitive Support

Piet Kommers

Abstract The motivation for this chapter is to show the evolutionary nature from earlier student tracking and 'intelligent tutoring systems' until the current stages of advanced learning analytics. This chapter's added value is its illustration of how learning analytics contributes to learners' meta-cognitive awareness and how indirectly a teachers' 'dashboard' allows teachers, parents, and remedial teachers to trace the more subtle cognitive entailments in the student's mind. The sources for the chosen direction rest upon earlier research into media-support for conceptual learning. This chapter may elicit the discussion on how far technology-driven innovations can bypass the existing repertoire of earlier didactic solutions.

Introduction

Learning analytics recently evolved based on big data, machine learning, deep learning, and data mining techniques. Starting from the utilitarian perspective the question is opportune how longer-term learning analytic profiles may contribute to the quality of teaching guidance and meta-cognitive awareness for the learner. In this chapter, the best candidates for integrating learning analytics outcomes will be presented and discussed for optimal adoption and dissemination in learning practices. Crucial starting point is what didactic paradigms have been chosen from the very start. The main antecedent of learning is still cognitive science. And, cognitive science is a hybrid amalgam as well: psychology, philosophy, linguistics, social, neuro and computer science. Learning analytics is not only relying on this six-pack; it also rests upon the more recent didactic innovation paradigms as formulated by Madhumita Bhattacharya in Heinrich et al., (2007); see Fig. 1.

The epitome of these five trends is 'active learning,' subordinated by the four paradigms: constructivist, cooperative, authentic and intentional learning. As teachers attempt to innovate didactic practices, these five are the main foci in order to make students more active learners. Learning analytics and its needed big data

---

P. Kommers (✉)
UNESCO Learning Technologies, University of Twente, Mooienhof 179, 7512EE Enschede, The Netherlands

E. McKay (ed.), *Manage Your Own Learning Analytics*, Smart Innovation, Systems and Technologies 261, https://doi.org/10.1007/978-3-030-86316-6_10

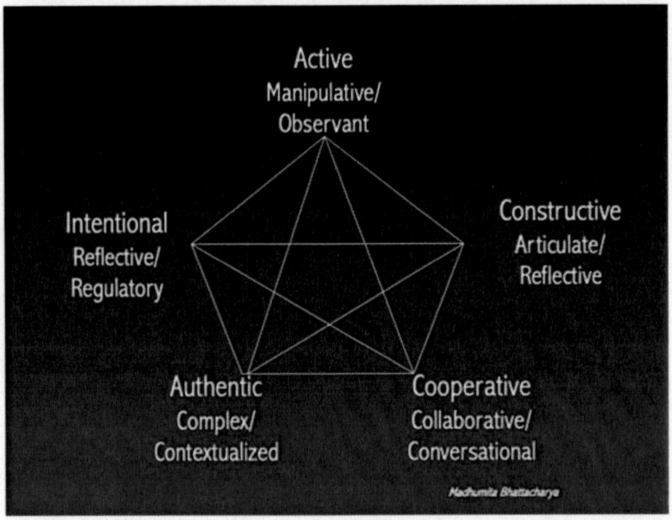

Fig. 1 Five dominant educational innovation paradigms by Madhumita Bhattacharya, 2007

are bound to the relevant parameters for each of these five dimensions. The next relevant shift is the trend toward integrating the work and playing methods in the learning arena. Through mobile learning, we see a swift trend to make learning ubiquitous. Typically, one hears the phrase: 'You can learn everywhere, even in school.' It completely mirrors the constant pressure to stimulate life-long learning as one of the twenty-first century skills and attitudes; learning has expanded its position from school- and course-based up to the level of a 'life-style.' If we want learning analytics to be successful, it is a matter of making our criteria of successful learning compatible with these new paradigmatic trends. A good example is the re-excavation of the 'vicarious learning' idea; it not only points us to the relevance of collaborative learning (employers need team players rather than soloists). It goes even one step further: Learners need to learn from its other's learning. Recently, problem-based learning has shown a remarkable affordance to vicarious learning; students who undertake complementary learning tasks indeed show a tendency to benefit from each other approaches and conclusions, provided that the instructional arrangement stimulates this and even enforces learners to rephrase the learning outcomes of their co-students.

One of the most pervasive shifts is to accept playing to be beneficial to learning and even to working; see Fig. 2. Traditionally, playing is seen as a prepatory stage to learning like in kindergarten. However, the gaming element has entered schools, vocational training, and even the medicine against burn-outs, and if we want to prepare citizens for the life-long learning phase after formal education, it is inevitable to allow playfulness in the classroom as otherwise the transfer from school to career will be hampered. Preliminary conclusion is that if we want learning analytics to be sensitive for the relevant parameters of learning success, we need to take all

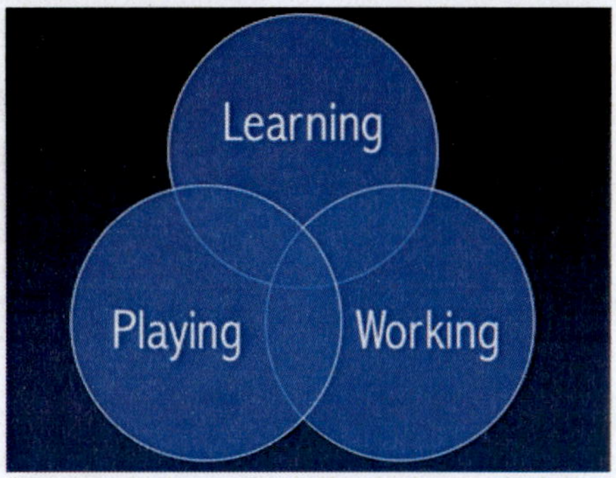

Fig. 2 Learning-playing-working triangulation

innovative trends into account. Merely, focusing on the traditional outcomes like memorization, comprehension, and textual communication ignores the new skills like critical thinking and collaborative problem solving. The underpinning computational methods for learning analytics can be seen as catalytic to its successful implementation: 'big data' and 'machine learning.'

Learning Analytics Stands on the Shoulders of Big Data, Deep and Machine Learning

The advent of advanced computing, data-analysis, and interconnecting databases has brought the ambition to support learning and teaching processes with 'smart' information. So far, the major attempts have been to increase the sophistication of instruction; a better diagnosis of the learning progress in the student, in order to adapt further guidance and feedback. The arrival of the 'constructivist' learning approach brought the option to make the student a better learner; knowledge was apprehended as the result of a delicate reconciliation between prior and new concepts, where the learner becomes and active orchestrator rather than a docile recipient. Learning analytics is the attempt to harvest ingredients from the student's past performance in order to optimize the evaluation and the future learning process as well. From this perspective, it is understandable that learning analytics has been welcomed as instructional support method. Good example is the 'dashboard' for teachers who can now see based upon learning analytics outcomes how to characterize an individual student in terms of strengths and weaknesses and even detect a certain learning style that can be used for tailoring further coaching and instruction. This chapter targets the complementary exploitation of learning analytic outcomes: How can the learner gain understanding on his/her trend in learning processes like the memory/understanding and top-down/bottom-up ratio. Crucial in the first-order benefits is the enrichment of conceptual representations; how can schematic topic entailments become more meaningful when past learning events are considered? In order to find generic answers, the triangularity between technology, the culture, and the nature of learning need to be clarified.

Learning Analytics: Evolutionary Step Forward?

The notion that learning can be arranged better when making use of past data has been expressed by learning psychologists like Hermann Ebbinghaus (1850—1909) and Edward Thorndike (1874–1949). The main paradigm became that learning need to be underpinned by prerequisite skills and knowledge. Since instructional strategies have defended the primacy of learning sequences and diagnostics for optimizing the learning process, the logged interaction data have become attractive assets to build upon. In order to enter the arena of pleading for certain analytic approaches, it will be good to discuss first the wider context of what we call: The tripod of educational technology and its evolution amidst the 'culture' and 'nature of learning' see Fig. 3.

As the dichotomy 'culture' versus 'nature' of learning is not well-known yet, and let us discuss its rational. Large parts of school education rests upon conventions like the need for 'concentration,' 'reducing sensory sensations other than the instructional channel,' and 'progressing from factual to procedural to higher order thinking.' In other words: 'Controlling the external conditions' has received optimal attention for the 'conditions of learning.' How different is the 'nature of learning'? It is the full spectrum of mental, physical, and emotional repertoire that emerge from our body and brains. Exactly, these unshakable factors like wet brains and hormones are becoming the two candidates for the longer term growth that allow us to develop self-organization, excavate long-term memories and the courage to leave our comfort zone. Quite convincing exemplars of nature factors in learning are: 1. The role of slow-wave sleep for consolidating impressions of the prior day. 2. Optimizing

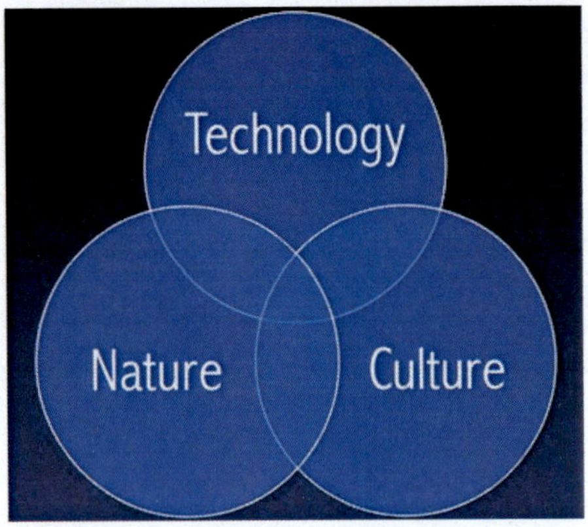

Fig. 3 Educational technology for arranging the optimal culture conventions and interventions in among the nature factors in learning

learning conditions via food extracts. 3. Utilizing mindfulness for bringing learners in a better balance for concentration and reflection. Learning analytics offers the option to allow instructional designers to take many more dimensions into account and find unthought ideal combinations. Just like we accept that teachers take decisions on the very moment based upon intuition, it will become common sense to let educational researchers open the plethora of Bayesian methods like the famous Demster—Shafer model. It allows researchers to combine evidence from different types of sources and reach at a more explicit level of expectations until a certain 'belief function' encompasses as many secondary and tertiary empirical sources as possible. Learning analytics are not just powerful representations for the sake of learning; they touch a highly sensitive aspect of privacy as well. Let us, therefore, look into recent political alertness.

Impact and Privacy of Learning Analytics

Learning analytics implies the collecting, rubricating, and reporting of student interaction data in their prior learning sessions. Its goal is to understand better what students go through, in order to optimize the further learning conditions and individual coaching. The 'dashboard' is this most common metaphor for teachers to interact with the data-analytic outcomes. The more controversial (mis)use is the exploitation of student data for external parties like future potential employers, and life insurance companies. That is why the Dutch parliament has enforced the Ministry of Education to limit publishers and courseware providers to utilize and fully make learning analytic data anonymous. For this reason, the 'low hanging fruit' for learning analytics has been seen in direct teacher support for instance to detect factors that predict student drop-out.

Learning Analytics for Personalizing Education

Crucial for conveying learning processes is the willingness to see patterns in human development. And, even more important is one's pedagogical mission; keen on helping someone who is midst of an existential process. And, this is not the entire story; also, the teacher is in a constant process of refining his/her didactic approaches. Analytics fits well in the larger tradition to personalize the teaching/learning relationship. Before starting the data-driven approach, it would be good that the teacher deepens the notion of learning and social/cognitive development. This is crucial as data mining rests upon the Bayesian approach of having an initial interest, and only then, querying the database can emerge. What are typical interests a teacher may have on an individual student's profile?

The first parameter is 'how the student is doing in terms of motivation, overall achievement, and typical deficiencies in certain domains of topics?' Interviews with teachers learnt that what they need most is the "out of school circumstances" like absence, complications in the home situation, medical factors, and embedding in the peer group. Its reason is easy: Teachers tend to focus on the pure didactic attributes of students; however, the social and pedagogical periphery may easily be overlooked. This means, however, that parents, social, and medical workers need to deliver global indications on a students' personal circumstance. Learning analytics focuses on detecting trends and patterns in large collections of educational data. It can provide descriptive, diagnostic, predictive, and prescriptive data. The first two face hindsight, the last two insight and foresight. The level of sophistication like the needed data mining and machine learning increases through this climax; however, it not necessarily increases the value of its outcomes. Before going to the large-scale trends and school interventions, it is key to assist teachers with the practical question on how to diagnose and remediate individual learners.

Ingredients for Learning Analytics

Machine learning (already coined in 1959) has been brought forward in domains where human expertise revealed mediocre solutions like in fuzzy decision making and games with an unknown depth of hidden rules like recently demonstrated in the AlphaGo program. Essentially, all available analysis rationales can be adopted: multivariate regressions, game theory, control theory, operations research, information theory, simulation-based optimization, multi-agent systems, swarm intelligence, statistics, and genetic algorithms. The interaction with human actors has stayed behind as machine learning derives its power from chaining infinite chains of permutations and unbridled sequences of induction and even seeds of randomness. For the sake of human–machine synergy, it will be hard to overestimate the value of transparency and explaining the immanent metacognition so that it allows the person to feel in sync with the track pattern by its machine thinking partner. This paper delivers the most salient architectures for the man/machine meta-cognitive interface.

Learning Paradigms

Large part of both man and machine learning is based upon bottom-up feature analysis like principal component analysis and sparse dictionary learning. After enduring efforts to formalize human learning, it becomes more and more clear that learning is one of the most versatile and diverse processes offered by nature. Ultimately, we can say that learning is 'the willingness to change oneself.' It opens the ontological question how change can be assessed by the subject itself? For the sake of machine learning, however, its existential approach offers a new dimension: What about an engine that has the potential to monitor and adapt its prior strategies and build new learning approaches on top of that? Meta-cognition has been recognized by renowned human learning theories as the best candidate for 'learning to learn' (Kommers, 1996).

Learning Analytics; Analogy to Medical Optimization

So far, most of the learning analytics interventions have stayed at the level of teacher support systems. The interface metaphor is usually conceived as a dashboard system that hides redundant complexity and articulates pragmatic indications for needed interventions. For instance, if the data show social complexities of a certain student, the teacher will include pedagogical and socio-emotive measures in his/her didactic repertoire. Machine learning for rendering the flow from surface symptoms to underlying pathological mechanisms and entailing patient care has a longer tradition based upon medical expert systems in the mid-eighties. The common interpretation of machine learning has been rather inert for the last twenty years, according to William Vorhies' blog post recently.

> *... The application of any computer-enabled algorithm that can be applied against a dataset to find a pattern in the data. This encompasses basically all types of data science algorithms, supervised, unsupervised, segmentation, classification, and regression including deep learning.*

Since BERT NLP models were first introduced by Google in 2018, they have become the go-to choice. New evidence, however, shows that LSTM models may widely outperform BERT meaning you may need to evaluate both approaches for your NLP project. In: https://www.datasciencecentral.com/profile/WilliamVorhies.

Figure 4 shows how in case of unsupervised learning the phase of clustering allows to bootstrap the next phase of improving practical interventions. Crucial for the success of adequate clustering is the integration of a wide spectrum of secondary variables like (in case of education) student and teacher appreciation. Traditionally, we would criticize this approach as 'grand fishing expedition.' In the era of big data, it becomes common practice to include secondary and tertiary data sources, in

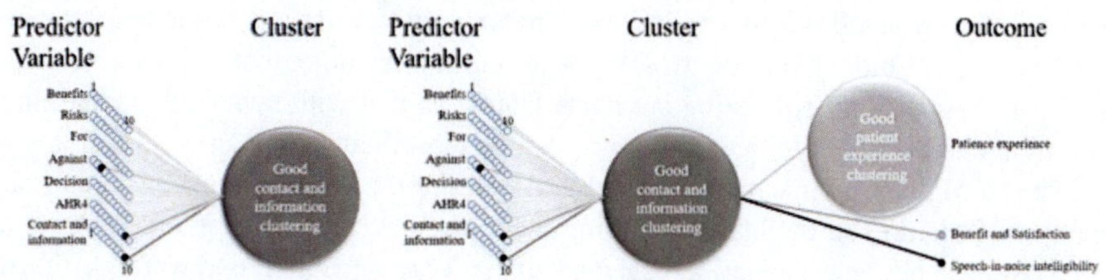

Fig. 4 Machine learning for clustering patients' symptoms

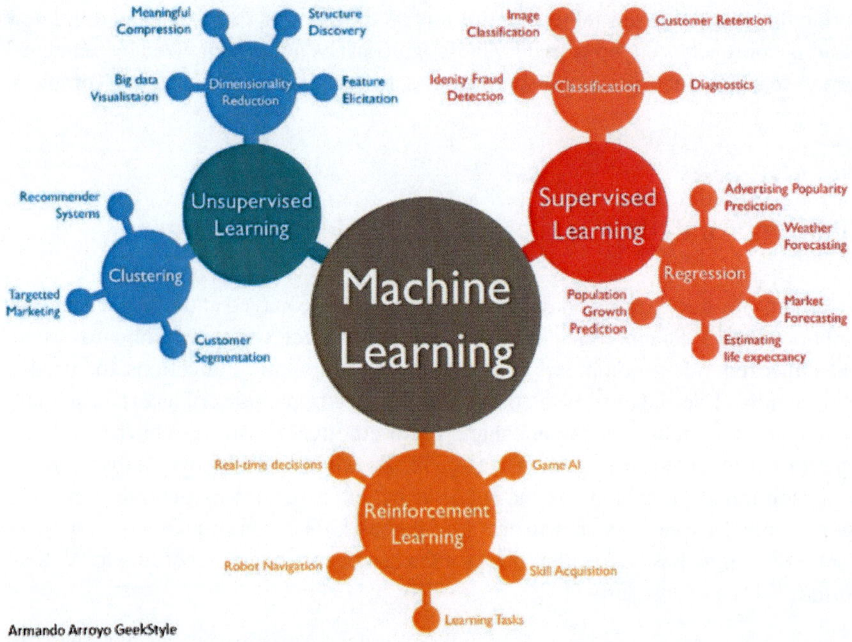

Fig. 5 Represented case studies about machine learning and predictive analytics as presented in the 'World for Business' Las Vegas, June 3–7, 2018

order to increase the serendipity effect; not finding what you looked for, however, unexpectedly finding relationships that are even more valuable for increasing the multi-facetted processes like education, teaching, and learning.

As Fig. 5 shows there are three main strands in the mechanisms underlying machine learning. The first one is deep learning; here called 'unsupervised learning' that rests upon reducing dimensionality. Easy example is when large datasets show that students' tendency toward 'intrinsic motivation' goes to a large extent together with 'parents' academic achievement' in combination with 'divergent thinking.' In this case, the factor 'in- versus extrinsic motivation' can be mapped back unto parental intellectual attitude and raises the predictive power on how a student develops in the transition from primary to secondary education.

The second one is 'supervised learning' in which covariates and intermediate variables are scouted in order to increase the explained variance as needed to make a deterministic model for 'controlling instructional arrangements.' Its underlying model is 'regression'; squeezing out noise factors so that with fewer data, a learning situation can become typified and trends can be classified.

The third class of machine learning is based upon reinforcement; those combinations of features in the learning setting that correlate with learning outcomes and effectiveness become labeled as 'successful' and are accepted as rewards for past performance and become promoted to '"favorable"' features of the learning setting.

It may be clear that the first one, 'unsupervised learning' has the highest potential as it crawls to the more powerful cluster of critical variables even without explicit hypothesizing of the researcher. In terms of learning analytics, the mechanism of 'unsupervised learning' seems to be a good candidate to tackle the first principles in new fuzzy constellations; for instance, when a totally new educational practice like game-, mobile-, or VR-based learning method emerges, and no traditional explanations fit the new ways of adopting and exploiting.

Meta-cognitive Representations for Learning Analytics

Whereas, machine learning mainly relies on mathematical optimization and reduction of dimensionality, human learning derives its power on utilizing the immense combinatorics of association space built through many interaction thoughts in our prior life episodes. The role of metacognition for building sustainable learning analytics systems is one of epistemic scaffolding; (Heeren & Kommers, 1992). For instance, as we can see in the diagram on 'concept mapping' in Fig. 6 below. The traditional role of concept mapping is to let students express prior and intermediate knowledge states. Once conceptual structures have been expressed and validated, they can be utilized for diagnosing learners' knowledge acquisition process.

Due to the structure in the reference graph (Fig. 6), some nodes with a high out degree like concept mapping, tools, implementations, and constructivism are

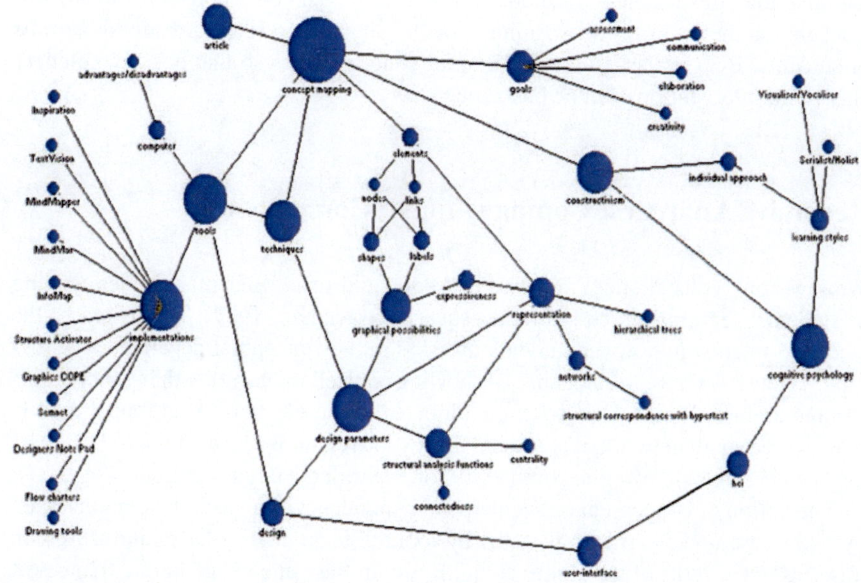

Fig. 6 Diagram on 'concept mapping' in knowledge transfer

displayed as nodes with a large diameter. It reflects the answer if we ask 'where to start the instructional sequence?' Learning analytics may consult the diagram with the questions in how far a student has focused on the 'central' versus the 'peripheral' concepts. Students with a linear cognitive style may show a direct trail from the start node until the far peripheral entailments of a certain branch; Huai & Kommers, 2000. Students with a holistic style may show breadth-first or even concentric patterns, reflecting elaborations; attempting to interlink the new to the prior concepts. In conclusion: Learning analytics goes beyond the administration of what Student A has been studying or not; it may reveal a typical tendency of how Student A tries to conquer the uncertainty of 'do I estimate to remember what I just addressed?' or 'in how far do I allow myself to dwell into details, before having the larger picture of the complete topic?' In summary: Meta-cognitive representations allow learning analytics to diagnose learning sequential patterns in order to help teachers to diagnose certain study deficits and its transformations along time. Once knowledge structures are represented, and learning analytics has the potential to make more complex inferences on students' styles, habits, anxiety failure, and even assumptions on students' motivation (intrinsic/extrinsic), ambition (the courage to explore topics that exceed the level of abstraction that typically fits to a certain level of schooling. Quite often, teachers and parents differ in opinion if a student suffers from one-sided high talents, concentration problems or just lack of interest. Also, here, learning analytics may offer teachers granular data on what exactly happened between the learner and the informational device, like nowadays the World Wide Web; Stoyanova & Kommers, 2001). In ongoing projects, we are exploring user interfaces, both for learners and teachers that allow to see sequential patterns in a glance. The chosen metaphor is the learning analytics teachers' dashboard. It expresses the continuous need to support teachers 'on the fly': only very limited time to get an impression on individual traits of students. But, just articulate meaningful characteristics so that effective plenary classroom interventions can be based on.

Recursive Analytics: Coping with NP Completeness

We saw how 'constellations' instead of 'sequential lists' offer more understanding in students' learning processes; (Levesque & Brachman, 1987). For instance, the teachers' interest in how far students invested in the conceptual dependency across topics. Networks have the elegance of showing constellations rather than strict orders and hierarchies. Figure 7 shows the undirected graph of intertwined topics A to F. Even with a small network, it is evident that cyclic relations like the loop C-D-B-A-C may cause some algorithmic complexity for a learning analytics engine.

The notion of conceptual or even topical dependency can easily be demonstrated by defining a node's structural status by looking to its first-order relationships. In this sense, the status of the Nodes A and B recruit the same value; both 2. However, including the Level 2 relationships reveal that Node A derives status from C and B, while Node B derives status from Nodes A and D. In order to obtain a quick checksum

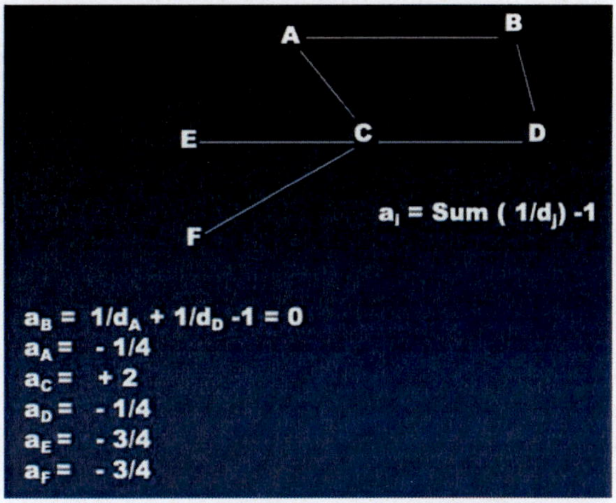

Fig. 7 Conceptual embeddedness in terms of centrality of nodes and topics

of the sum of all node statuses equaling 0, the definition of centrality status can be defined as the sum of reciprocal values of its first-order neighbors. Compliant to the visual perception, Node C has the best structural position (value equals + 2) to act as hub to all other nodes. Node B has a neutral position (equals 0). All other nodes have a negative attraction: (Nodes A and D equaling—one-fourth), while both Nodes E and F have the structural disposition of—three-fourth. The analogion of conceptual representation with learning diagnostics is 'recursive dependency': Knowledge elements derive its understanding from its subordinate relatives and these again from its subordinates, etc. Summary: Learning analytics for the domain of knowledge acquisition, metacognition, and structural understanding need to take the wider constellation of conceptual constellations into account. Even for cyclic networks (acquiring a new concept contributes to its own further understanding).

Mapping for Meta-cognitive Imagination

Both students, teachers and educational administrators have a potential need for learning analytic outcomes. The first expectation is that more sophisticated representations offer better diagnostics during and after the learning process evolves. Its merit is a wider set of monitoring tools and the potential of remedial and differentiation actions. At the same time parliaments, parents and school organizations have warned that these detailed student process profiles have a high degree of privacy and potential threat for being misused in the hands of employers, life insurance prospects, etc. This led researchers to start with a complete anonymization before any analyses take place.

The second expectation is the learning analytic data may become crucial for the sake of the student him/herself. A good example is the concept map in Fig. 6.

Fig. 8 Transition from the London underground map to connection circuit

It allows students to develop a meta-cognitive view on the browsing tour before and the imagination on how to proceed a next time. The big potential is the lucid formats that analytics can offer for student–student communication and student–teacher negotiations. The advantage of the second one is that students stay owners of their own learning processes; the teacher may look into it; however, its interpretation develops in the dialog between the two of them.

Toward Didactic Topologies for Analytics

Various representational methods will be considered the coming decades, for the sake of making the learning process more transparent, lucid, and humane. A good example is the transformation from the strict isomorphic map into the circuit diagram as we may see in the evolution of the London underground map, in Fig. 8.

We may expect learning analytic dashboards to arrive the coming years that do not simply attempt to display learning parameters, however, more and more toward guidelines that should help teachers and students to see navigational recommendations. Most important is that learning analytics need more explicit learning and didactic paradigms. In this sense, learning analytics will prove to be catalytic for coming innovations, rather than just collecting a plethora of data.

References

Heeren, E. & Kommers, P. (1992). Flexibility of expressiveness: A critical factor in the design of concept mapping tools for learning. In P. Kommers, D.H. Jonassen, & J.T. Mayes (Eds.), *Cognitive tools for learning: Proceedings of the NATO advanced study institute on mindtools: cognitive technologies for modeling knowledge*, July 4–10 1990 (pp. 85–102).

Heinrich, E., Bhattacharya, M., & Rayudu, R. (2007). Preparation for lifelong learning using ePortfolios. *European Journal of Engineering Education, 32*(6), 653–663. https://doi.org/10.1080/03043790701520602

Huai, H. & Kommers, P. (2000). *Concept mapping as a learning strategy for serialist students*. Paper presented at 16 World Computer Congress, WCC 2000.

Kommers, P. (1996). Hypermedia and multimedia concepts: Definitions. In P. A. M. Kommers, S. Grabinger, & J. C. Dunlap (Eds.), *Hypermedia learning environments: Instructional design and integration* (pp. 1–12). Lawrence Erlbaum.

Levesque, H. & Brachman, R.J. (1987). Expressiveness and tractability in knowledge representation and reasoning. *Computational Intelligence*, *3*(1).

Stoyanova, N., & Kommers, P. (2001). Learning effectiveness of concept mapping in a computer-supported collaborative problem-solving design. In P. Dillenbourg, A. Eurelings, & K. Hakkarainen (Eds.), *European perspectives on computer-supported collaborative learning: proceedings of the First European conference on computer-supported collaborative learning:* March 22–24, 2001, Universiteit Maastricht, the Netherlands (pp. 561–568). Unigraphic.

Piet Kommers is an early pioneer in media for cognitive- and social support. His doctoral research explored methods for hypertext and concept mapping in learning. Since 1982 he developed educational technology for teacher training. His main thesis is that technology is catalytic for human ambition and awareness. His main function is associate professor in the University in Twente, The Netherlands and adjunct/visiting professor in various countries. He taught more than fifteen bachelor-master- and Ph.D. courses and supervised more than 30 Ph.D. students. He instigated and coordinated the NATO Advanced Research Workshop on Cognitive Technologies in 1990 and a large series of Joint European Research Projects in: authoring multimedia, web-based learning, teacher education, virtual 3d worlds, constructivist learning, social media, web-based communities and international student exchange. UNESCO awarded his work in ICT for Education in Eastern Europe with the title of Honorary Professor. The Capital Normal University in Beijing awarded his work with the title of Honorary Doctor. He is member of advisory boards in ministries of education and academia of sciences in Singapore, Finland and Russia. Piet Kommers is the initiator of the international journal for web-based communities and overall chair of the IADIS conferences on societal applications of ICT. Since the late nineties he gave more than 40 invited and keynote lectures at main conferences in the fields of education, media and communication. His books and journal articles address the social and intellectual transformations at each transition from 'traditional' into the 'new' media. Instead of regarding media as extrapolating, supplanting, vicarious or even disruptive, Piet's view is that new media elicit and seduce both individuals and organizations to reconsider human nature and challenge existential awareness at that very moment. His main publications and citation score can be found here. His workshop templates and experiences have been implemented into the UNESCO IITE reports, policy briefings and Master Course. The books and journal articles of Piet Kommers reach the level of 5012 citations and the h-index of 30.ResearcherID: B-8673–2014 Scopus Author ID: 55723206300